My Book

This book belongs to

Name: _____

1

www.math-knots.com

www.math-knots.com

Cover Design by :
Gowri Vemuri

First Edition :
May, 2019

Author :
Gowri Vemuri

Edited by :
Raksha Pothapragada
Ritvik Pothapragada

Questions: mathknots.help@gmail.com

NOTE : VDOE is neither affiliated nor sponsors or endorses this product.

Dedication

This book is dedicated to:

My Mom, who is my best critic, guide and supporter.

To what I am today, and what I am going to become tomorrow,

is all because of your blessings, unconditional affection and support.

This book is dedicated to the

strongest women of my life ,

my dearest mom

and

to all those moms in this universe.

G.V.

A **mathematical** operation performed on a numbers in order to divide it equally is called division. **Division** is one of the four basic operations of arithmetic, the others being addition , subtraction and multiplication. Several symbols are used for the division operator, including the obelus (÷), the (-) and the slash (/).

Division is splitting into equal parts or groups. It is also referred as "fair sharing".

Division is the **opposite of multiplying**.

There are special names for each number in a division.

dividend ÷ divisor = quotient

But Sometimes It does not work perfectly!

Sometimes we cannot divide things up exactly ... there may be something left over.

dividend = divisor X quotient + remainder

Dividend: Any number that needs to get divided into equal shares (No of shares equal to divisor)

Divisor: The number of equal shares the dividend is getting divided.

Quotient: It is the number that says how much each share gets.

 www.math-knots.com

INDEX

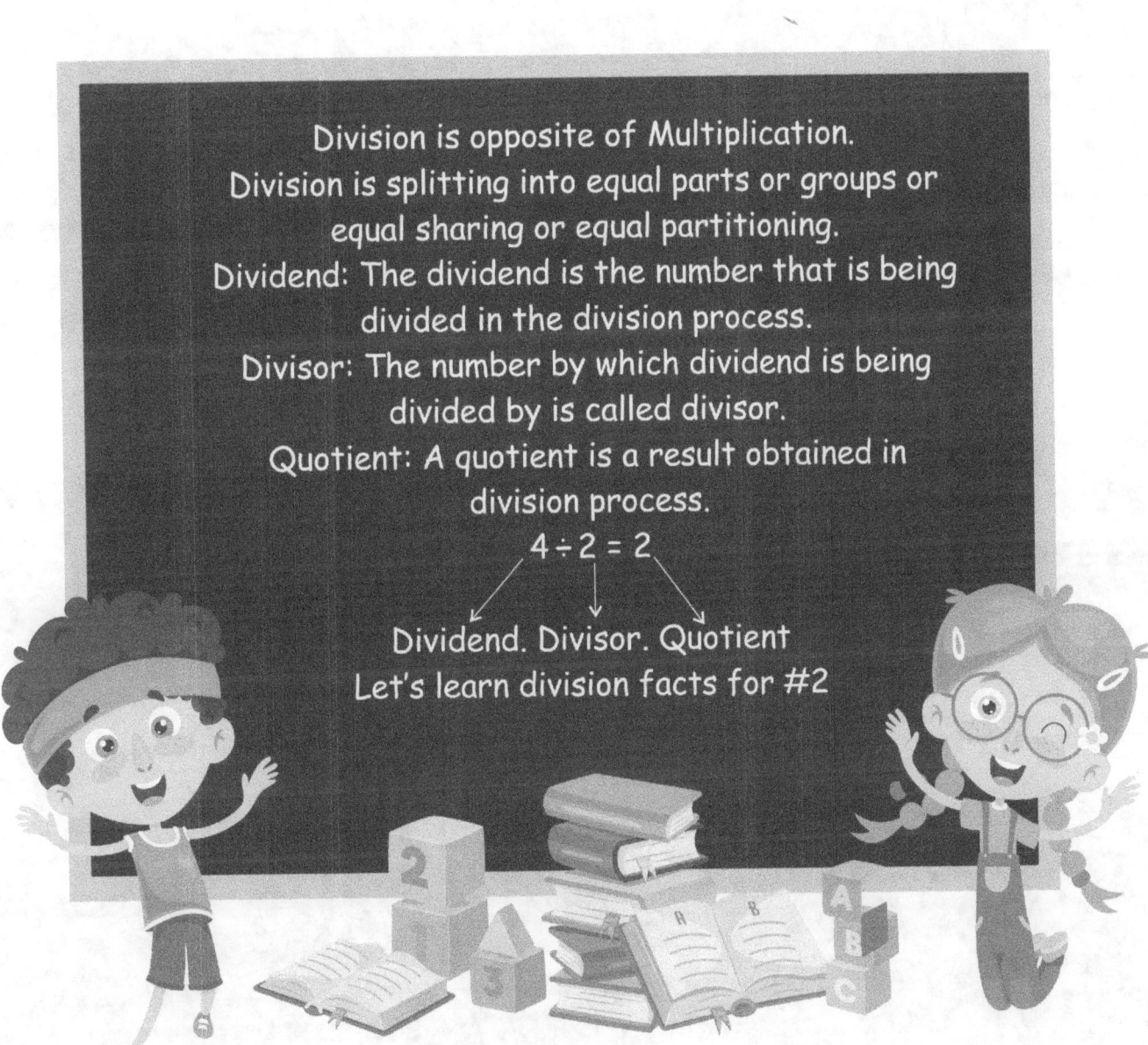

Division is opposite of Multiplication.
Division is splitting into equal parts or groups or equal sharing or equal partitioning.
Dividend: The dividend is the number that is being divided in the division process.
Divisor: The number by which dividend is being divided by is called divisor.
Quotient: A quotient is a result obtained in division process.

$$4 \div 2 = 2$$

Dividend. Divisor. Quotient
Let's learn division facts for #2

www.math-knots.com

www.math-knots.com

1. Lets learn $2 \div 1 = 2$

A.

B.

C. $\boxed{2 \div 1 = 2}$

2. Lets learn $4 \div 2 = 2$

A.

B.

C. $\boxed{4 \div 2 = 2}$

3. Lets learn $6 \div 2 = 3$

A.

B.

C. $\boxed{6 \div 2 = 3}$

4.   Lets learn $8 \div 2 = 4$

A.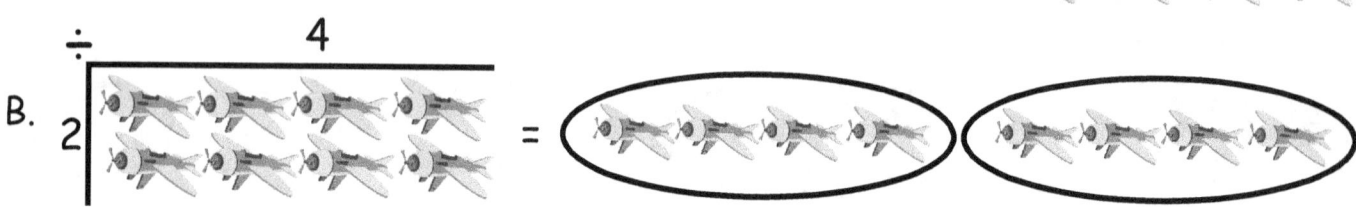

B. $2\overline{)}$ 4

=

C. $\boxed{8 \div 2 = 4}$

5.   Lets learn $10 \div 2 = 5$

A.

B. $2\overline{)}$ 5

=

C. $\boxed{10 \div 2 = 5}$

6.   Lets learn $12 \div 2 = 6$

A.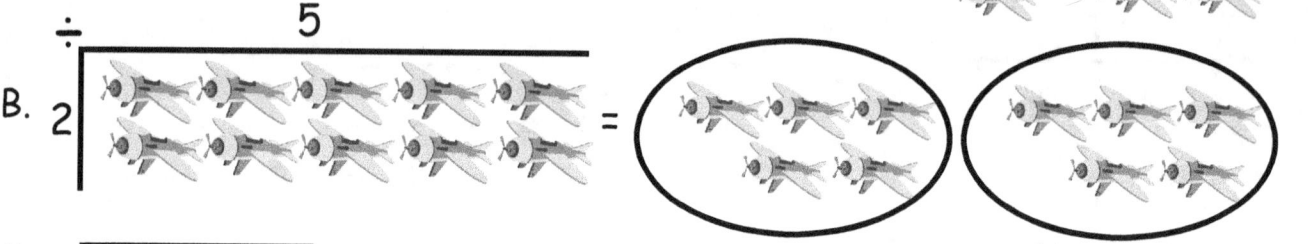

B. $2\overline{)}$ 6

=

C. $\boxed{12 \div 2 = 6}$

www.math-knots.com

7.   Lets learn 14 ÷ 2 = 7

A.

B.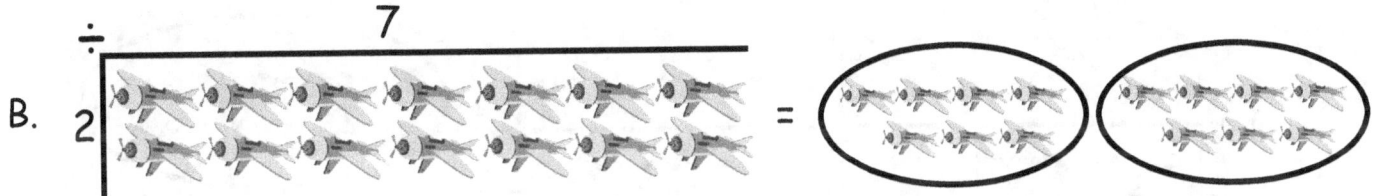

C. $\boxed{14 \div 2 = 7}$

8.   Lets learn 16 ÷ 2 = 8

A.

B.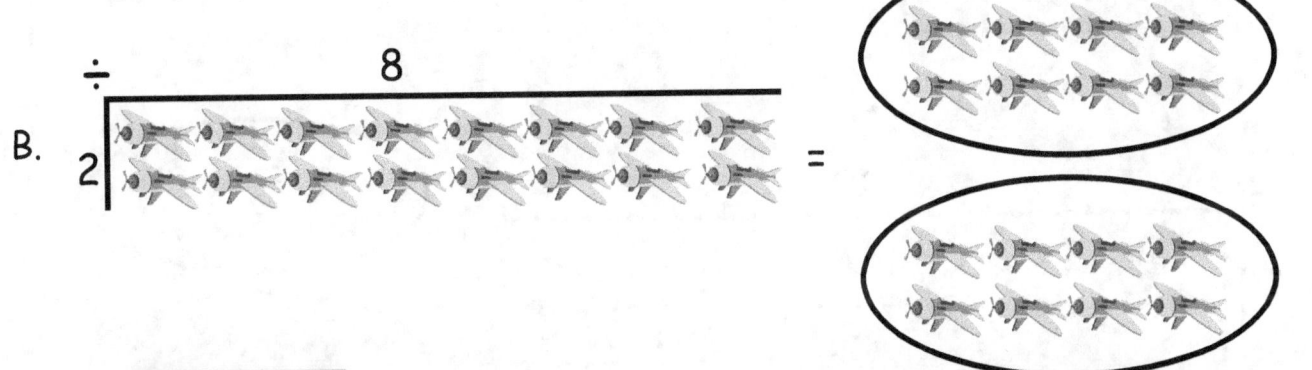

C. $\boxed{16 \div 2 = 8}$

9. Lets learn 18 ÷ 2 = 9

A.

B. ÷
    9
    2 [fish] =
    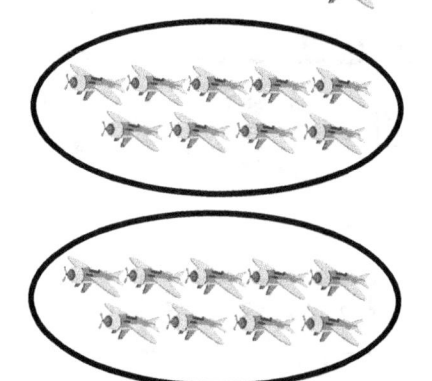

C. **18 ÷ 2 = 9**

10. Lets learn 20 ÷ 2 = 10

A. [fish diagram]

B. ÷
    10
    2 [fish] =
    [fish circles]

C. **20 ÷ 2 = 10**

11. Lets learn 22 ÷ 2 = 11

A.

B.

$$\frac{11}{2}$$

C. | 22 ÷ 2 = 11 |

Did you know division by 2 means
Dividing the given number into 2 equal share's ?
Dividing the number into two equal Groups.

www.math-knots.com

12. Lets learn 24 ÷ 2 = 12

A.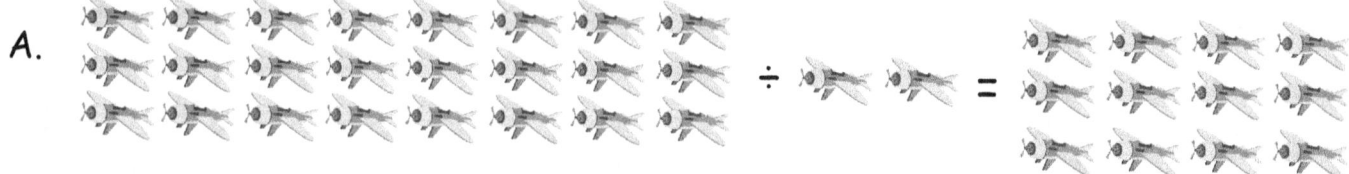

B.

C. 24 ÷ 2 = 12

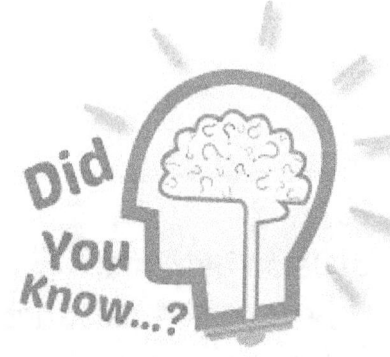

Did You Know...?

Did you know you can write division sign
in three different ways
÷ , / and ──

www.math-knots.com

Exercise - 1

(A) $2\overline{)2}$ (F) $2\overline{)12}$ (K) $2\overline{)22}$

(B) $2\overline{)4}$ (G) $2\overline{)14}$ (L) $2\overline{)24}$

(C) $2\overline{)6}$ (H) $2\overline{)16}$ (M) $2\overline{)26}$

(D) $2\overline{)8}$ (I) $2\overline{)18}$ (N) $2\overline{)28}$

(E) $2\overline{)10}$ (J) $2\overline{)20}$ (O) $2\overline{)30}$

www.math-knots.com

Exercise - 2

1.	2 ÷ 2 =	_____
2.	4 ÷ 2 =	_____
3.	6 ÷ 2 =	_____
4.	8 ÷ 2 =	_____
5.	10 ÷ 2 =	_____
6.	12 ÷ 2 =	_____
7.	14 ÷ 2 =	_____
8.	16 ÷ 2 =	_____
9.	18 ÷ 2 =	_____
10.	20 ÷ 2 =	_____
11.	22 ÷ 2 =	_____
12.	24 ÷ 2 =	_____

1	× _____ =	2
2	× _____ =	4
3	× _____ =	6
4	× _____ =	8
5	× _____ =	10
6	× _____ =	12
7	× _____ =	14
8	× _____ =	16
9	× _____ =	18
10	× _____ =	20
11	× _____ =	22
12	× _____ =	24

Did You Know...?

Did you know division is splitting a number up by any give number.

www.math-knots.com

Exercise - 3

1. I am a number, I divide myself, into one equal group of 2. What am I ?

    (A) 0   (B) 1

    (C) 2   (D) 3

2. I am a number, I divide myself, into two equal groups of 1. What am I ?

    (A) 1   (B) 4

    (C) 6   (D) 2

3. I am a number, I divide myself, into two equal groups of 2. What am I ?

    (A) 6   (B) 8

    (C) 2   (D) 4

4. I am a number, I divide myself, into two equal groups of 3. What am I ?

    (A) 6   (B) 3

    (C) 8   (D) 2

5. I am a number, I divide myself, into two equal groups of 4. What am I ?

    (A) 4   (B) 6

    (C) 8   (D) 10

www.math-knots.com

6. I am a number, I divide myself, into two equal groups of 5. What am I ?

(A) 12   (B) 10

(C) 6   (D) 5

7. I am a number, I divide myself, into two equal groups of 6. What am I ?

(A) 6   (B) 2

(C) 14   (D) 12

8. I am a number, I divide myself, into two equal groups of 7. What am I ?

(A) 2   (B) 10

(C) 14   (D) 7

9. I am a number, I divide myself, into two equal groups of 8. What am I ?

(A) 16   (B) 8

(C) 20   (D) 14

10. I am a number, I divide myself, into two equal groups of 9. What am I ?

(A) 9   (B) 18

(C) 20   (D) 14

www.math-knots.com

11. I am a number, I divide myself, into two equal groups of 10. What am I ?

(A) 2   (B) 22

(C) 10   (D) 20

12. I am a number, I divide myself, into two equal groups of 11. What am I ?

(A) 20   (B) 22

(C) 11   (D) 24

13. I am a number, I divide myself, into two equal groups of 12. What am I ?

(A) 12   (B) 18

(C) 24   (D) 26

14. I am a number, I divide myself, into two equal groups of 13. What am I ?

(A) 26   (B) 18

(C) 20   (D) 13

15. I am a number, I divide myself, into two equal groups of 14. What am I ?

(A) 14   (B) 18

(C) 22   (D) 28

23

Exercise - 4

Solve the maze run below.

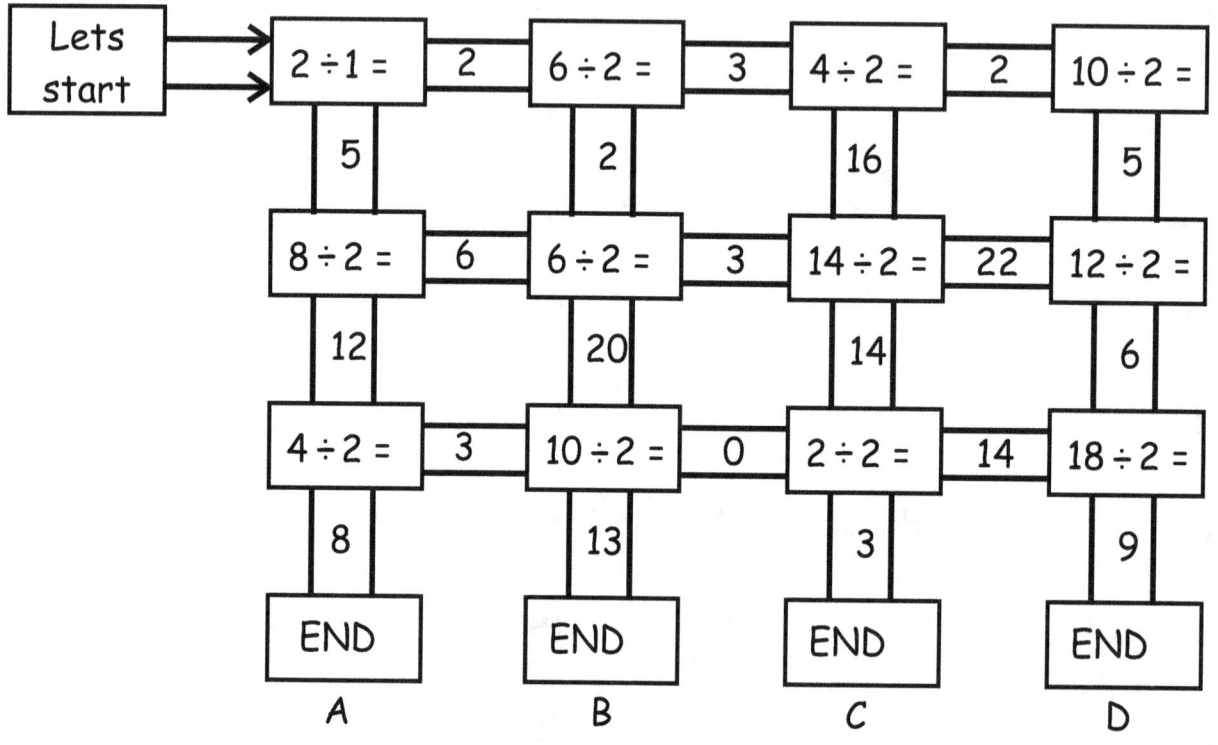

Lets start	2 ÷ 1 =	2	6 ÷ 2 =	3	4 ÷ 2 =	2	10 ÷ 2 =
	5		2		16		5
	8 ÷ 2 =	6	6 ÷ 2 =	3	14 ÷ 2 =	22	12 ÷ 2 =
	12		20		14		6
	4 ÷ 2 =	3	10 ÷ 2 =	0	2 ÷ 2 =	14	18 ÷ 2 =
	8		13		3		9
	END		END		END		END
	A		B		C		D

Who won the race ? _____

www.math-knots.com

Exercise - 5

1. 2 ÷ ☐ = 1   then ☐ = _____

2. 4 ÷ ☐ = 2   then ☐ = _____

3. 6 ÷ ☐ = 2   then ☐ = _____

4. 8 ÷ ☐ = 2   then ☐ = _____

5. 10 ÷ ☐ = 2   then ☐ = _____

6. 12 ÷ ☐ = 2   then ☐ = _____

7. 14 ÷ ☐ = 2   then ☐ = _____

8. 16 ÷ ☐ = 2   then ☐ = _____

9. 18 ÷ ☐ = 2   then ☐ = _____

10. 20 ÷ ☐ = 2   then ☐ = _____

11. 22 ÷ ☐ = 2   then ☐ = _____

12. 24 ÷ ☐ = 2   then ☐ = _____

Hey you are an expert of division facts of 2!!!

www.math-knots.com

Division is opposite of Multiplication.
Division is splitting into equal parts or groups or equal sharing or equal partitioning.
Dividend: The dividend is the number that is being divided in the division process.
Divisor: The number by which dividend is being divided by is called divisor.
Quotient: A quotient is a result obtained in division process.

$$6 \div 3 = 2$$

Dividend. Divisor. Quotient
Let's learn division facts for #3

www.math-knots.com

1. Lets learn 3 ÷ 1 = 3

A.  ÷  =

B.
$\div$ 3

1 | = (teddy bears circled)

C. | **3 ÷ 1 = 3** |

2. Lets learn 6 ÷ 3 = 2

A.  ÷  = (two teddy bears)

B.
$\div$ 2

3 | =  (three circles of two bears each)

C. | **6 ÷ 3 = 2** |

www.math-knots.com

3.   Lets learn 9 ÷ 3 = 3

A.

B.

3

$\div \quad 3$

=

C. $9 \div 3 = 3$

4.   Lets learn 12 ÷ 3 = 4

A.

B.

3

$\div \quad 4$

=

C. $12 \div 3 = 4$

5. Lets learn $15 \div 3 = 5$

A.

B.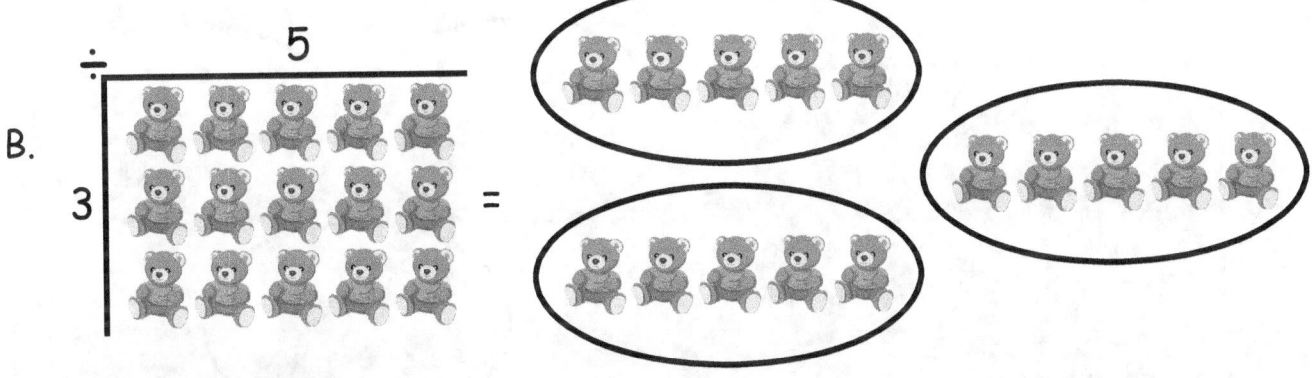

C. $\boxed{15 \div 3 = 5}$

6. Lets learn $18 \div 3 = 6$

A.

B.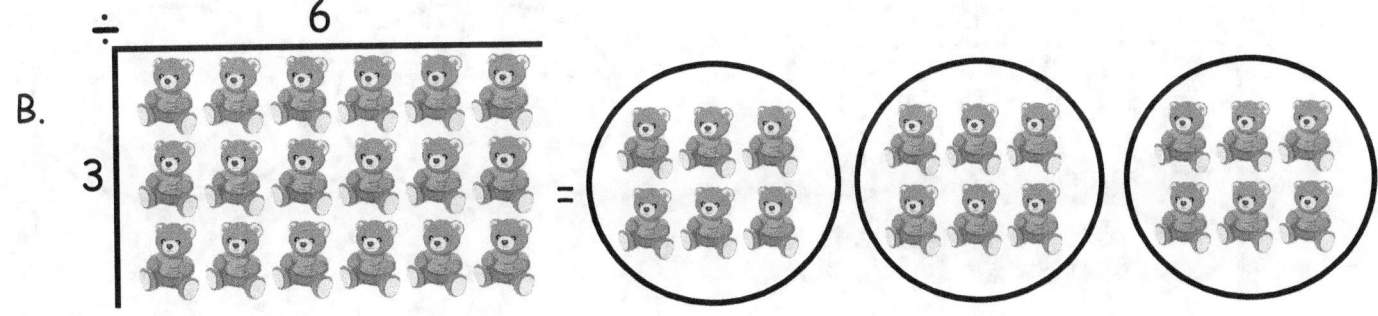

C. $\boxed{18 \div 3 = 6}$

www.math-knots.com

7. Lets learn $21 \div 3 = 7$

A.

B.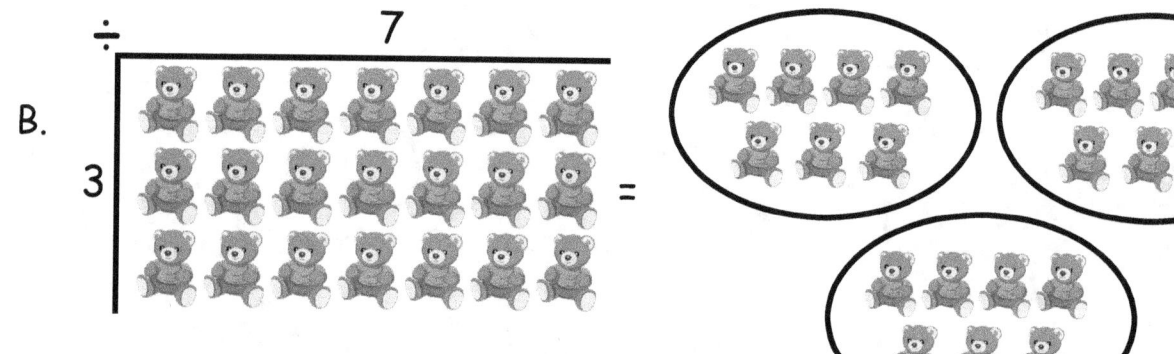

C. $\boxed{21 \div 3 = 7}$

8. Lets learn $24 \div 3 = 8$

A.

B.

C. $\boxed{24 \div 3 = 8}$

9.   Lets learn 27 ÷ 3 = 9

A.

B.

C.   27 ÷ 3 = 9

10.   Lets learn 30 ÷ 3 = 10

A.

B.

C.   30 ÷ 3 = 10

                   www.math-knots.com

11.　Lets learn 33 ÷ 3 = 11

A.
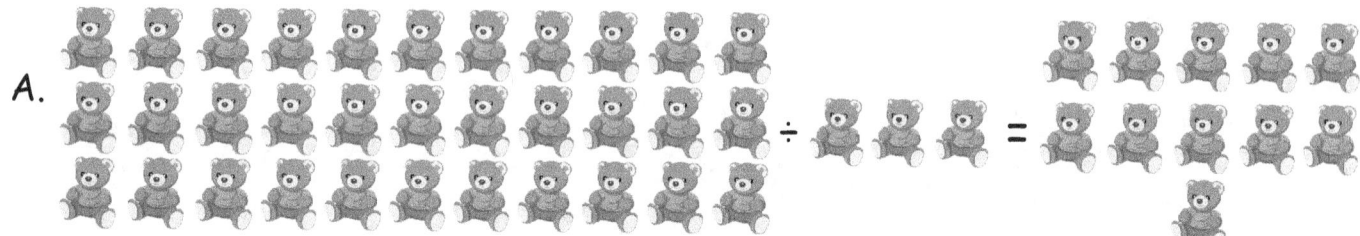

B.
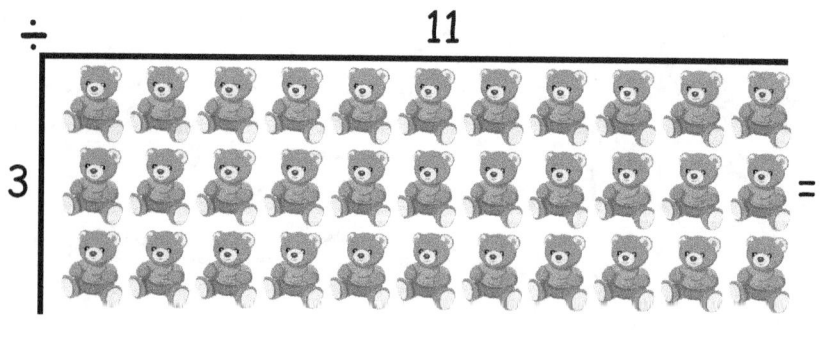

C.　| 33 ÷ 3 = 11 |

Did you know division by 3 means
Dividing the given number into 3 equal share's ?
Dividing the number into three equal Groups.

www.math-knots.com

12. Lets learn 36 ÷ 3 = 12

A.

B. ÷

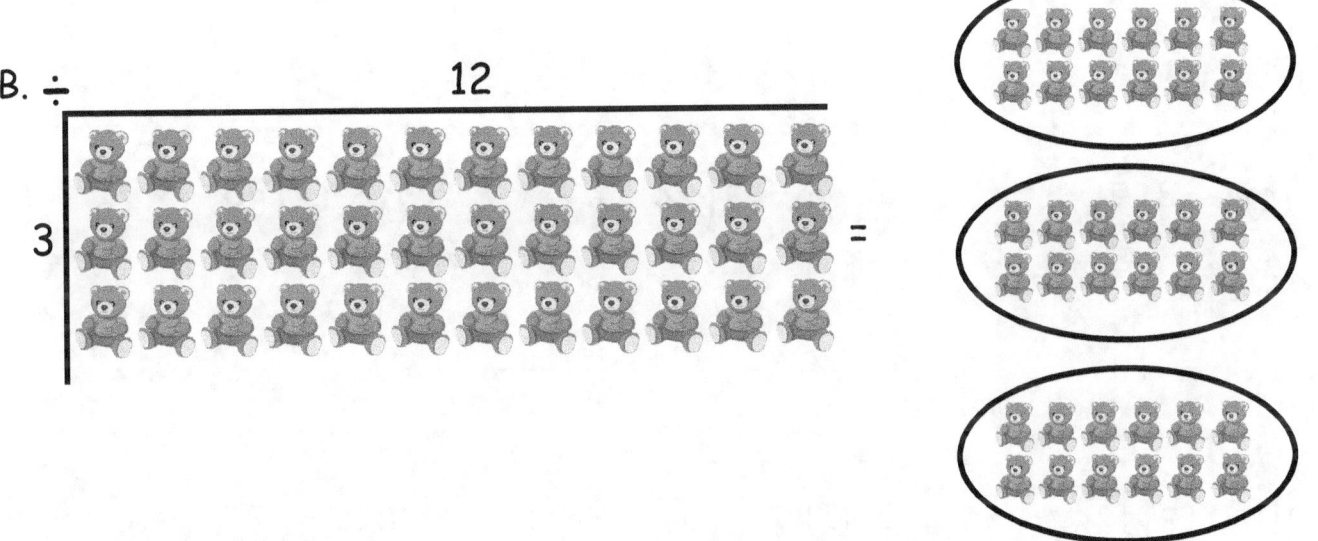

12

3

C. | 36 ÷ 3 = 12 |

Did you know you can write division sign
in three different ways
÷ , / and —

35 www.math-knots.com

Exercise - 1

(A) $3\overline{)3}$

(B) $3\overline{)6}$

(C) $3\overline{)9}$

(D) $3\overline{)12}$

(E) $3\overline{)15}$

(F) $3\overline{)18}$

(G) $3\overline{)21}$

(H) $3\overline{)24}$

(I) $3\overline{)27}$

(J) $3\overline{)30}$

(K) $3\overline{)33}$

(L) $3\overline{)36}$

(M) $3\overline{)39}$

(N) $3\overline{)42}$

(O) $3\overline{)45}$

www.math-knots.com

Exercise - 2

1.	$3 \div 3 =$ _____	$1 \times$ ____ $= 3$
2.	$6 \div 3 =$ _____	$2 \times$ ____ $= 6$
3.	$9 \div 3 =$ _____	$3 \times$ ____ $= 9$
4.	$12 \div 3 =$ _____	$4 \times$ ____ $= 12$
5.	$15 \div 3 =$ _____	$5 \times$ ____ $= 15$
6.	$18 \div 3 =$ _____	$6 \times$ ____ $= 18$
7.	$21 \div 3 =$ _____	$7 \times$ ____ $= 21$
8.	$24 \div 3 =$ _____	$8 \times$ ____ $= 24$
9.	$27 \div 3 =$ _____	$9 \times$ ____ $= 27$
10.	$30 \div 3 =$ _____	$10 \times$ ____ $= 30$
11.	$33 \div 3 =$ _____	$11 \times$ ____ $= 33$
12.	$36 \div 3 =$ _____	$12 \times$ ____ $= 36$

Did
You
Know...?

Did you know division is splitting a
number up by any give number.

 # Exercise - 3

1. I am a number, I divide myself, into one equal group of 3. What am I ?

   (A) 0   (B) 1

   (C) 2   (D) 3

2. I am a number, I divide myself, into four equal groups of 1. What am I ?

   (A) 1   (B) 4

   (C) 6   (D) 3

3. I am a number, I divide myself, into three equal groups of 2. What am I ?

   (A) 6   (B) 8

   (C) 2   (D) 4

4. I am a number, I divide myself, into three equal groups of 3. What am I ?

   (A) 6   (B) 3

   (C) 9   (D) 2

5. I am a number, I divide myself, into three equal groups of 4. What am I ?

   (A) 3   (B) 12

   (C) 4   (D) 10

www.math-knots.com

6. I am a number, I divide myself, into three equal groups of 5. What am I ?

(A) 12   (B) 9

(C) 15   (D) 5

7. I am a number, I divide myself, into three equal groups of 6. What am I ?

(A) 18   (B) 15

(C) 6   (D) 12

8. I am a number, I divide myself, into three equal groups of 7. What am I ?

(A) 7   (B) 21

(C) 3   (D) 15

9. I am a number, I divide myself, into three equal groups of 8. What am I ?

(A) 8   (B) 3

(C) 21   (D) 24

10. I am a number, I divide myself, into three equal groups of 9. What am I ?

(A) 18   (B) 21

(C) 27   (D) 9

39

11. I am a number, I divide myself, into three equal groups of 10. What am I ?

    (A) 10   (B) 30

    (C) 3   (D) 21

12. I am a number, I divide myself, into three equal groups of 11. What am I ?

    (A) 33   (B) 11

    (C) 3   (D) 27

13. I am a number, I divide myself, into three equal groups of 12. What am I ?

    (A) 30   (B) 33

    (C) 36   (D) 12

14. I am a number, I divide myself, into three equal groups of 13. What am I ?

    (A) 39   (B) 18

    (C) 21   (D) 13

15. I am a number, I divide myself, into three equal groups of 14. What am I ?

    (A) 14   (B) 33

    (C) 27   (D) 42

Exercise - 4

Solve the maze run below.

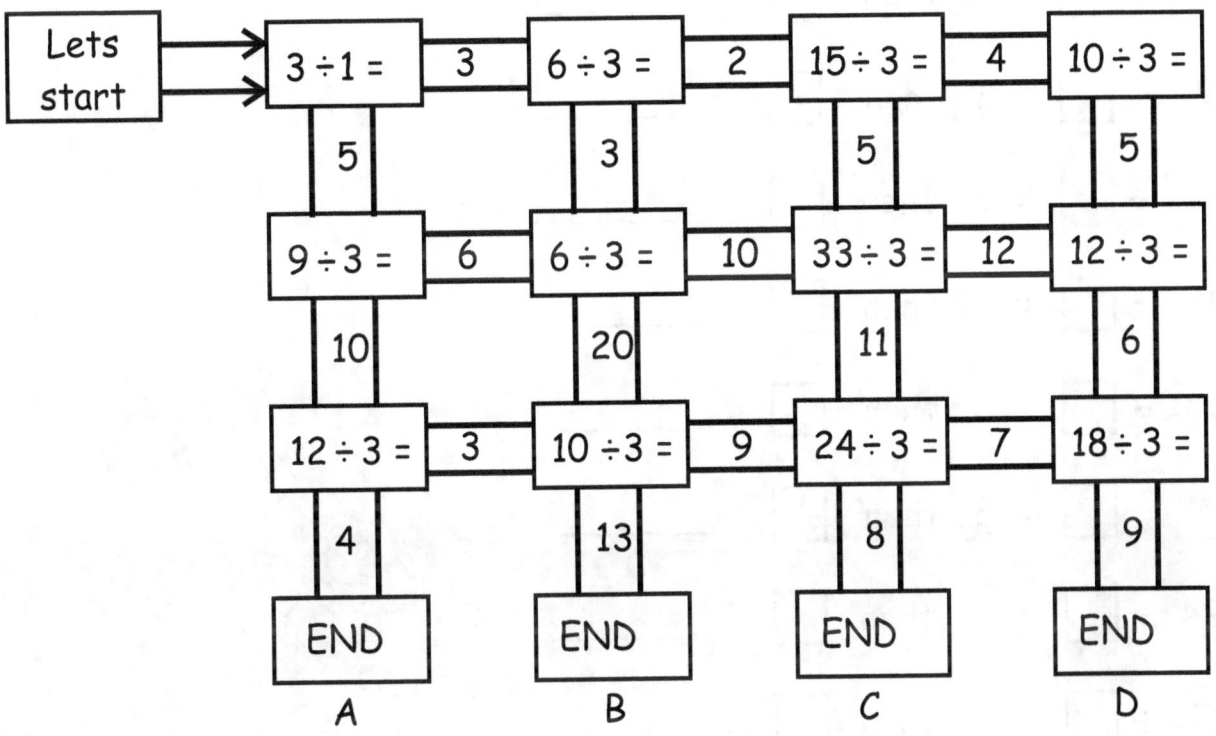

Lets start	3 ÷ 1 = 3	6 ÷ 3 = 2	15 ÷ 3 = 4	10 ÷ 3 =
	5	3	5	5
	9 ÷ 3 = 6	6 ÷ 3 = 10	33 ÷ 3 = 12	12 ÷ 3 =
	10	20	11	6
	12 ÷ 3 = 3	10 ÷ 3 = 9	24 ÷ 3 = 7	18 ÷ 3 =
	4	13	8	9
	END	END	END	END
	A	B	C	D

Who won the race ? _____

www.math-knots.com

Exercise - 5

1. $3 ÷ \boxed{} = 1$ then $\boxed{} =$ _____

2. $6 ÷ \boxed{} = 3$ then $\boxed{} =$ _____

3. $9 ÷ \boxed{} = 3$ then $\boxed{} =$ _____

4. $12 ÷ \boxed{} = 3$ then $\boxed{} =$ _____

5. $15 ÷ \boxed{} = 3$ then $\boxed{} =$ _____

6. $18 ÷ \boxed{} = 3$ then $\boxed{} =$ _____

7. $21 ÷ \boxed{} = 3$ then $\boxed{} =$ _____

8. $24 ÷ \boxed{} = 3$ then $\boxed{} =$ _____

9. $27 ÷ \boxed{} = 3$ then $\boxed{} =$ _____

10. $30 ÷ \boxed{} = 3$ then $\boxed{} =$ _____

11. $33 ÷ \boxed{} = 3$ then $\boxed{} =$ _____

12. $36 ÷ \boxed{} = 3$ then $\boxed{} =$ _____

Hey you are an expert of division facts of #3 !!!

www.math-knots.com

Division is opposite of Multiplication.
Division is splitting into equal parts or groups or equal sharing or equal partitioning.

Dividend: The dividend is the number that is being divided in the division process.

Divisor: The number by which dividend is being divided by is called divisor.

Quotient: A quotient is a result obtained in division process.

$$8 \div 4 = 2$$

Dividend. Divisor. Quotient

Let's learn division facts for #4

www.math-knots.com

www.math-knots.com

1. Lets learn $4 \div 1 = 4$

A.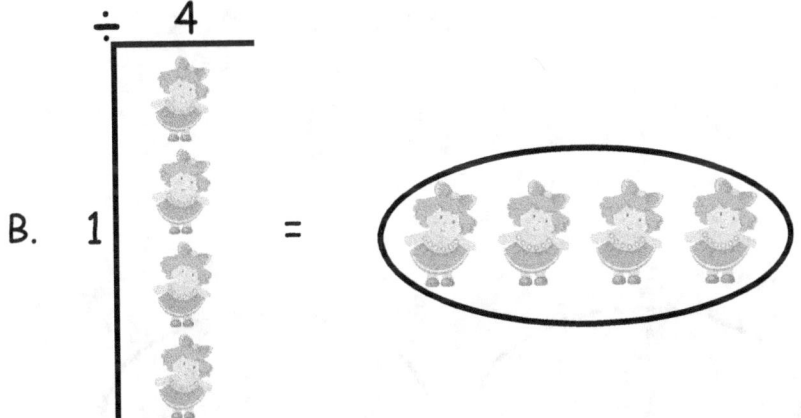

B.

C. $4 \div 1 = 4$

2. Lets learn $8 \div 4 = 2$

A.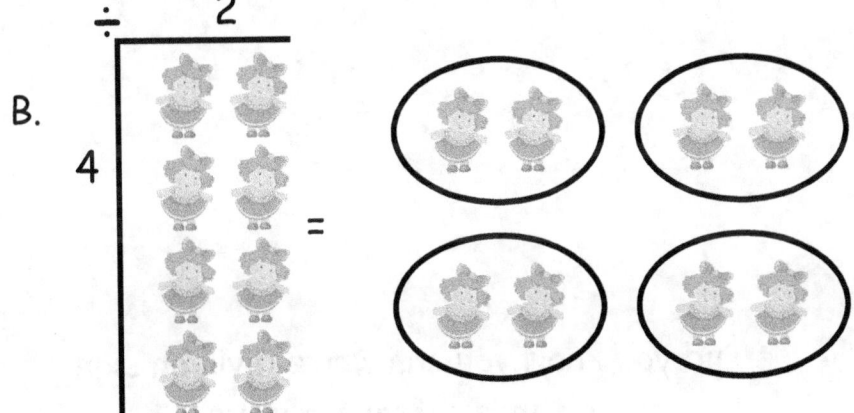

B.

C. $8 \div 4 = 2$

3.   Lets learn $4 \div 3 = 12$

A.    $\div$   =

B. $\div$ 3

   4   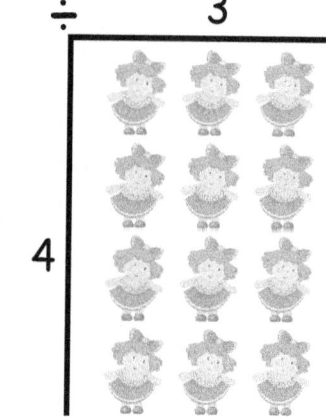   =

C. $12 \div 4 = 3$

Did you know you can write division sign
in three different ways
$\div$, / and —

www.math-knots.com

4. Lets learn 16 ÷ 4 = 4

A.  ÷

B. 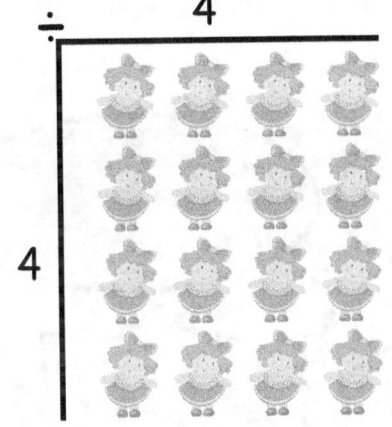 =

C. $\boxed{16 \div 4 = 4}$

Did you know division by 4 means
Dividing the given number into 4 equal share's ?
Dividing the number into four equal Groups.

www.math-knots.com

5. Lets learn 20 ÷ 4 = 5

A.

B.

÷
5
4

C. $20 \div 4 = 5$

Did you know division is splitting a number up by any give number.

www.math-knots.com

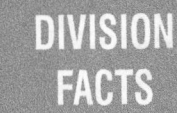

6.   Lets learn $24 \div 4 = 6$

A.

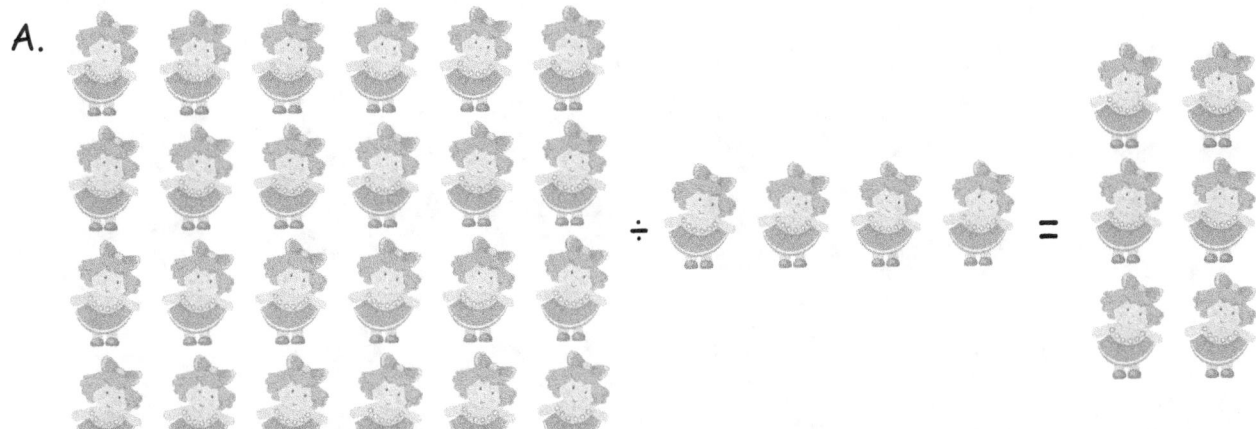

B.

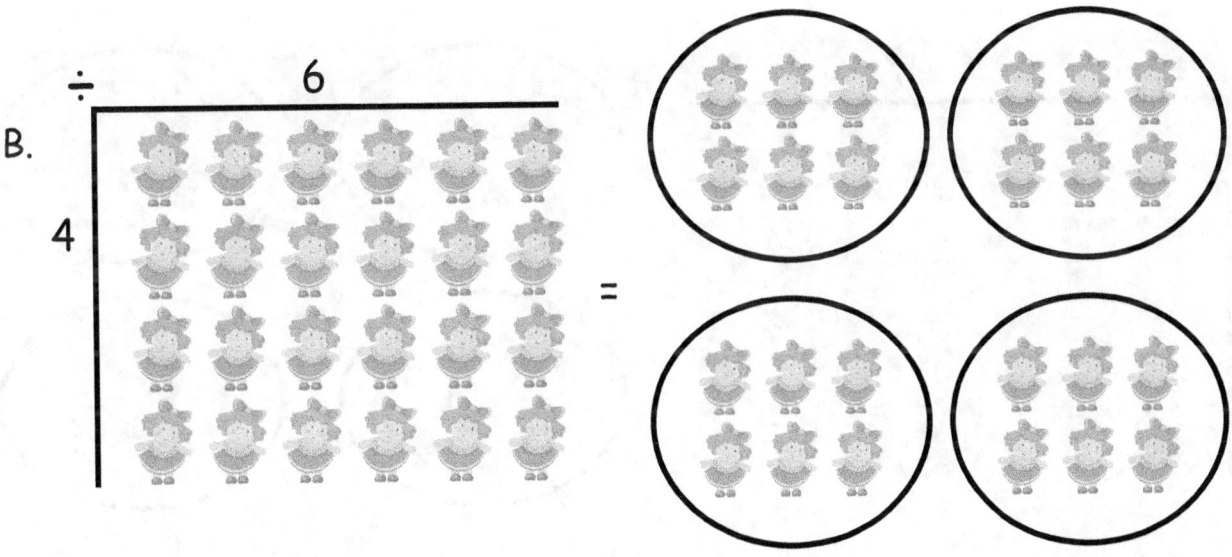

C. | $24 \div 4 = 6$ |

7.   Lets learn $28 \div 4 = 7$

A.

B.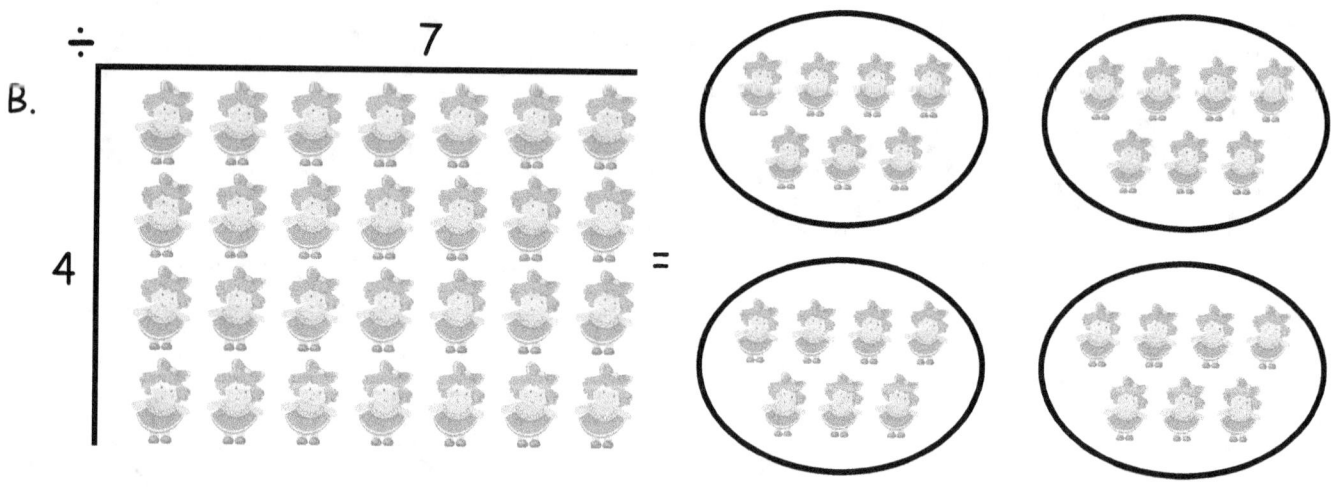

C. $\boxed{28 \div 4 = 7}$

www.math-knots.com

8.　Lets learn $32 \div 4 = 8$

A.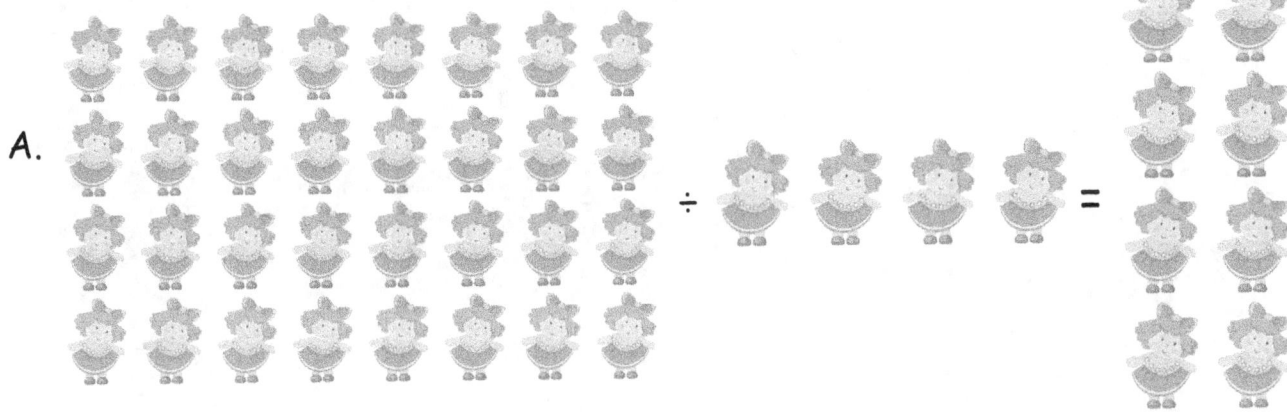

B.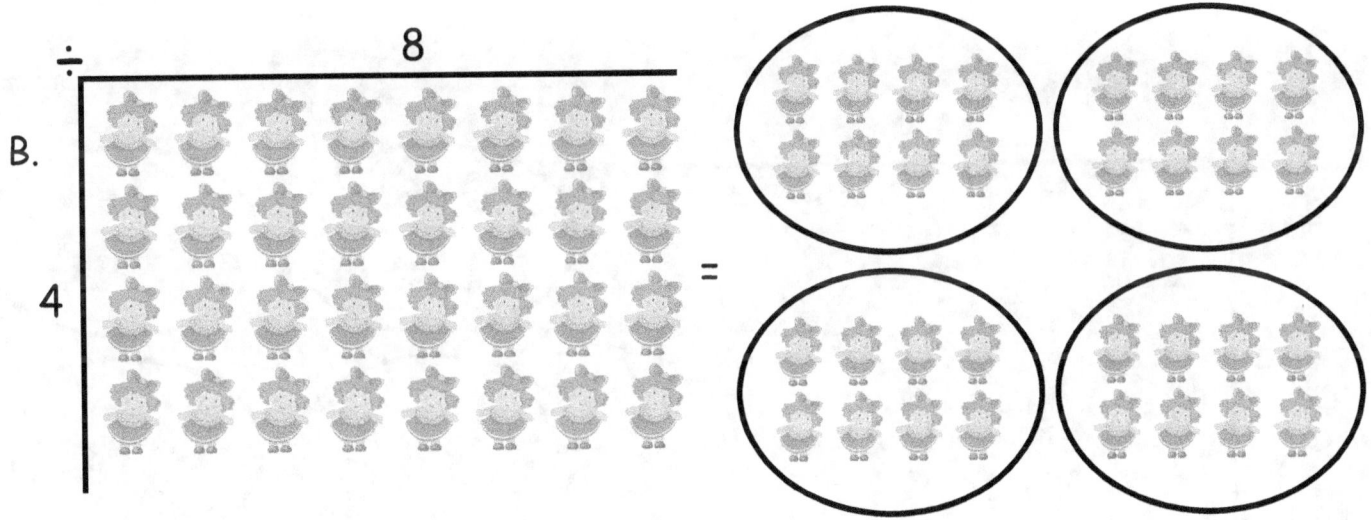

C.　| $32 \div 4 = 8$ |

9. Lets learn 36 ÷ 4 = 9

A.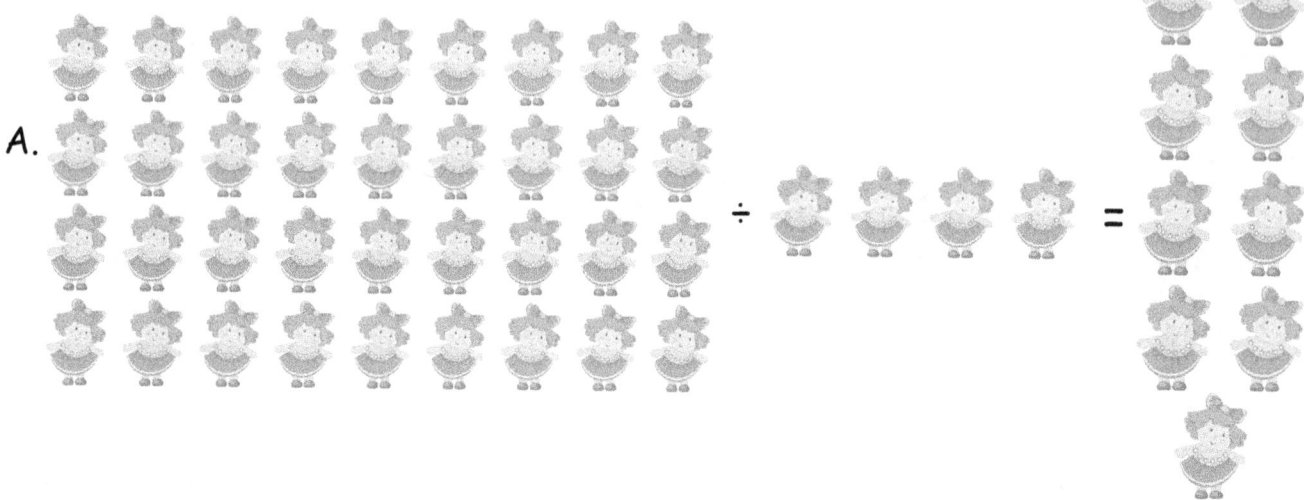

B.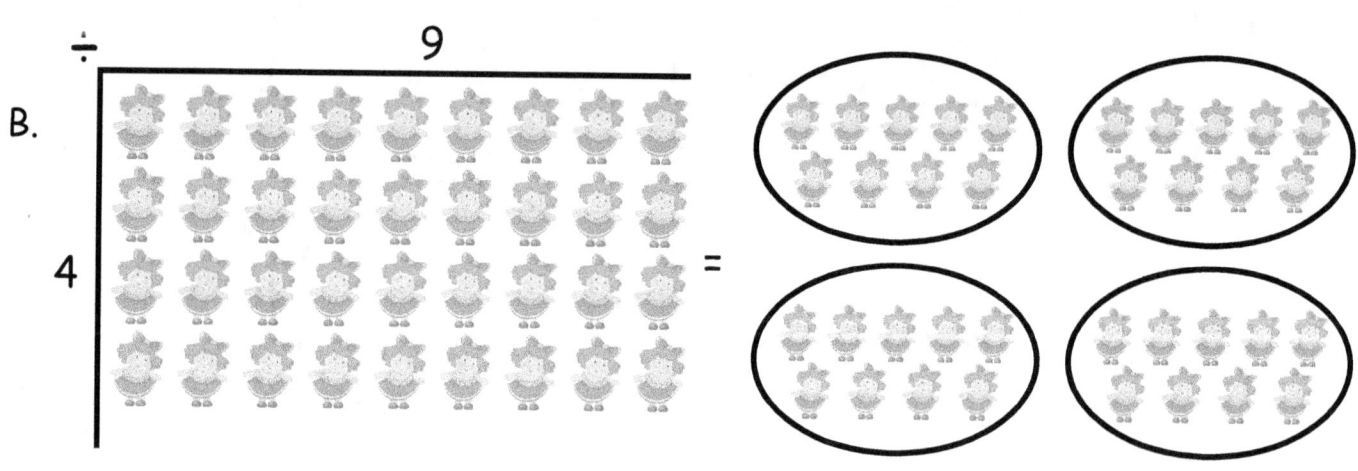

C.
$$36 \div 4 = 9$$

10.   Lets learn $40 \div 4 = 10$

A.

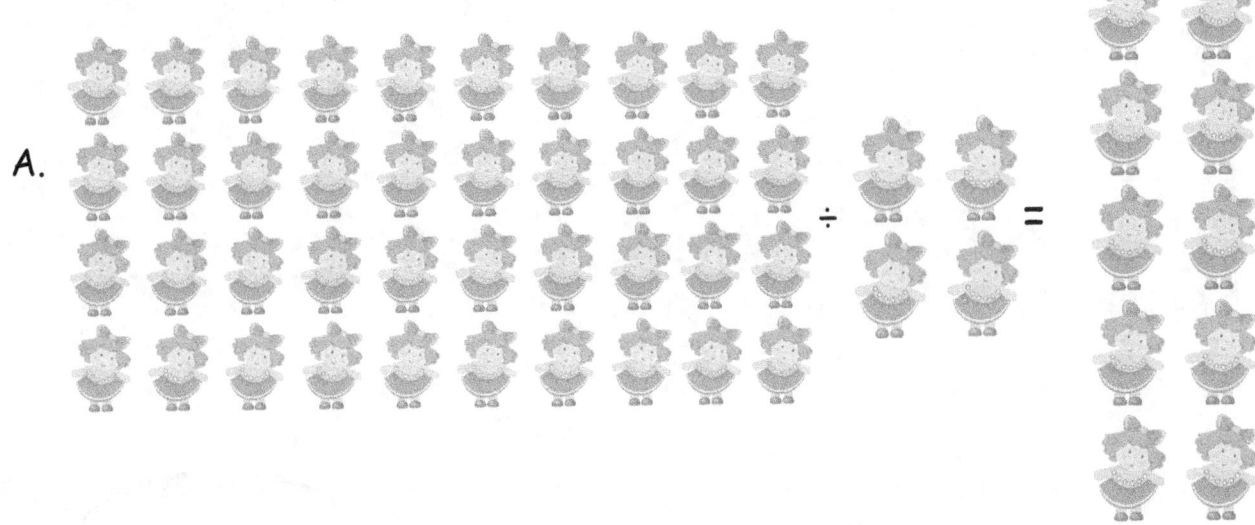

B.

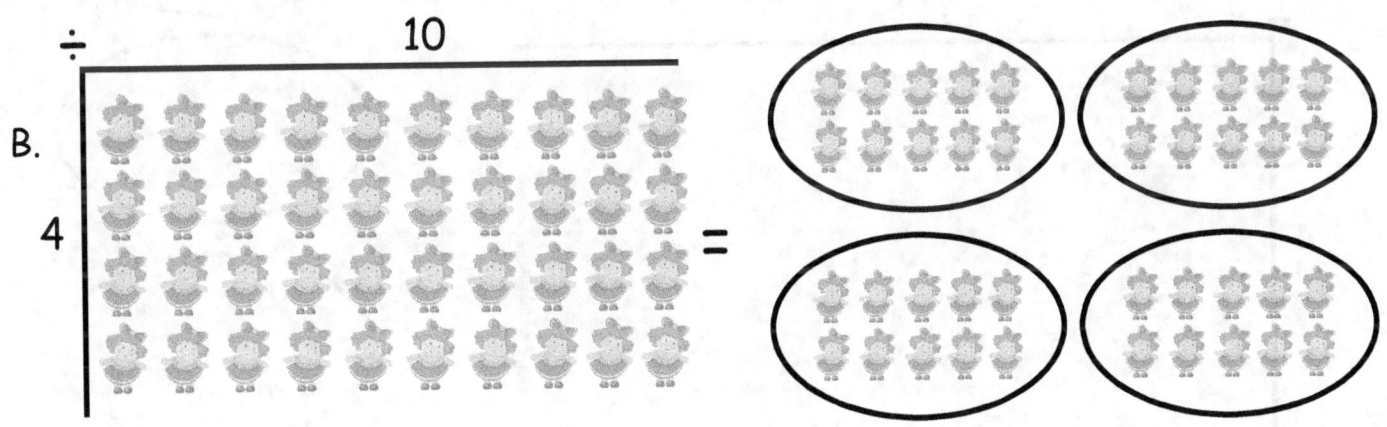

C. $\boxed{40 \div 4 = 10}$

11. Lets learn $44 \div 4 = 11$

A.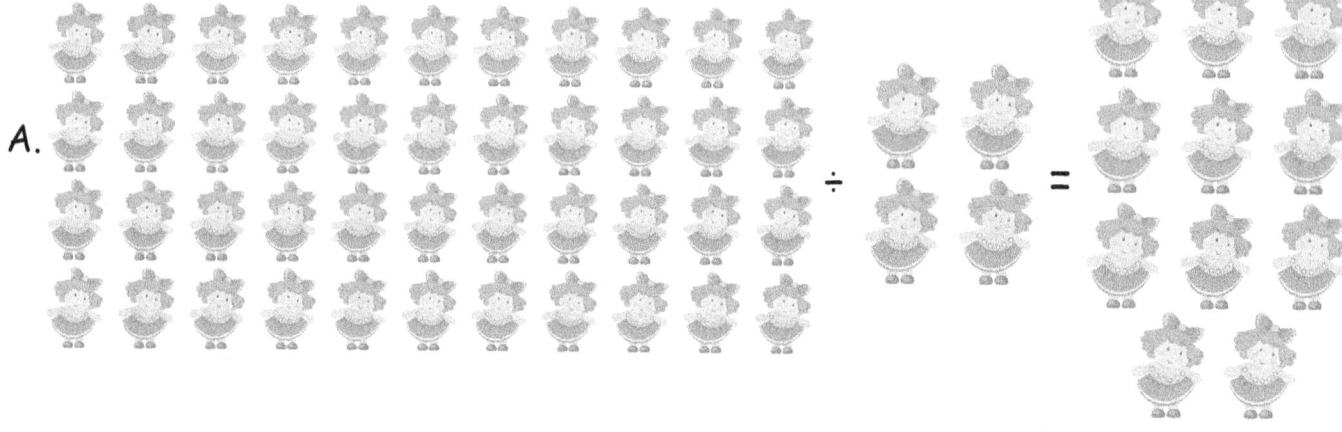

B.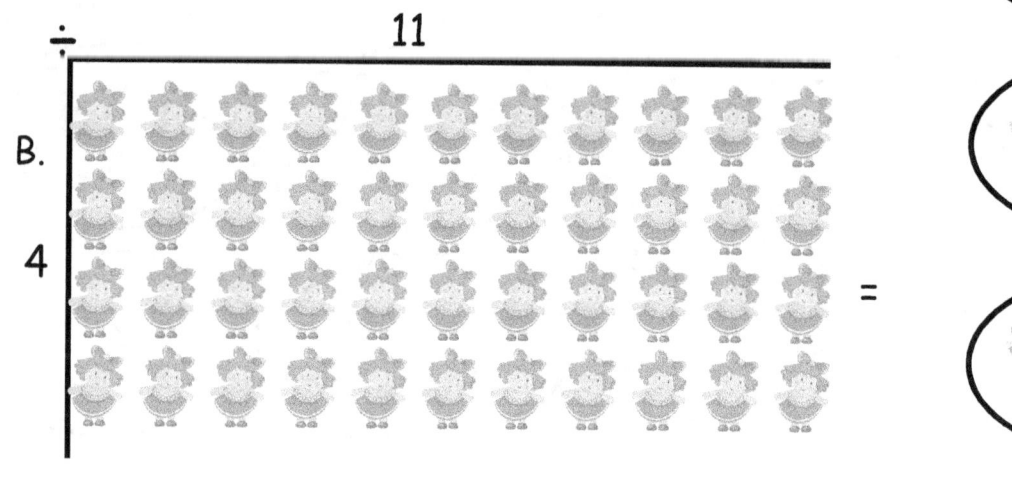

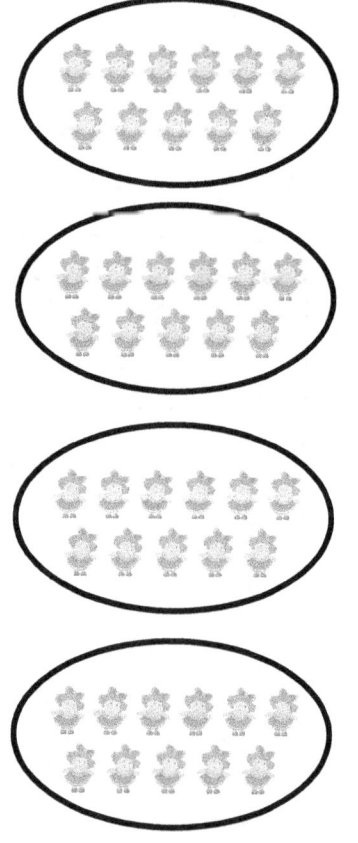

C. $\boxed{44 \div 4 = 11}$

12. Lets learn $48 \div 4 = 12$

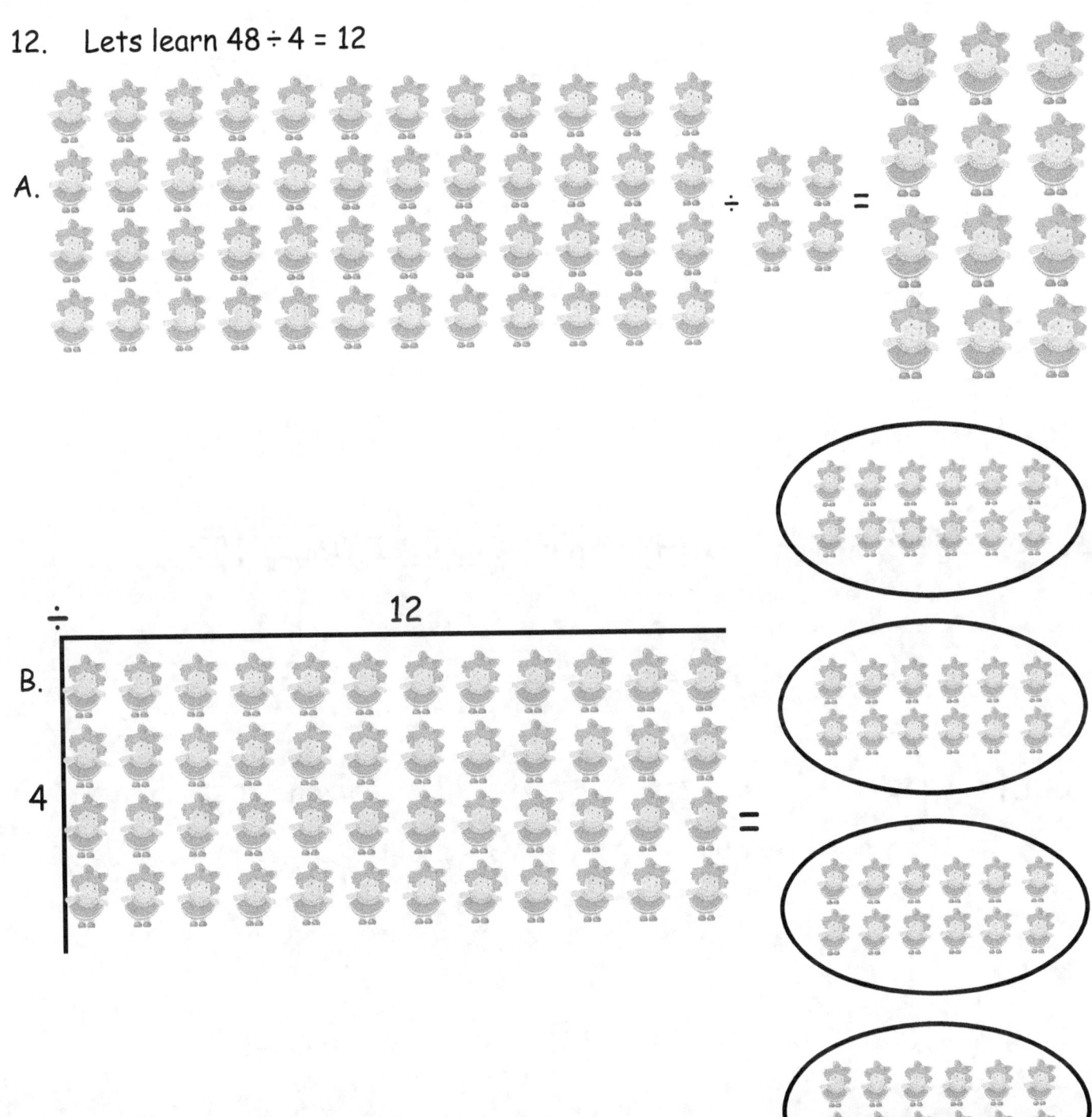

A.

B.

C. $48 \div 4 = 12$

Exercise - 1

(A) $4\overline{)4}$ (F) $4\overline{)24}$ (K) $4\overline{)44}$

(B) $4\overline{)8}$ (G) $4\overline{)28}$ (L) $4\overline{)48}$

(C) $4\overline{)12}$ (H) $4\overline{)32}$ (M) $4\overline{)52}$

(D) $4\overline{)16}$ (I) $4\overline{)36}$ (N) $4\overline{)56}$

(E) $4\overline{)20}$ (J) $4\overline{)40}$ (O) $4\overline{)60}$

www.math-knots.com

Exercise - 2

1.	4 ÷ 4 = _____
2.	8 ÷ 4 = _____
3.	12 ÷ 4 = _____
4.	16 ÷ 4 = _____
5.	20 ÷ 4 = _____
6.	24 ÷ 4 = _____
7.	28 ÷ 4 = _____
8.	32 ÷ 4 = _____
9.	36 ÷ 4 = _____
10.	40 ÷ 4 = _____
11.	44 ÷ 4 = _____
12.	48 ÷ 4 = _____

1	× _____ = 4
2	× _____ = 8
3	× _____ = 12
4	× _____ = 16
5	× _____ = 20
6	× _____ = 24
7	× _____ = 28
8	× _____ = 32
9	× _____ = 36
10	× _____ = 40
11	× _____ = 44
12	× _____ = 48

Did you know division is splitting a number up by any give number.

www.math-knots.com

Exercise - 3

1.   I am a number, I divide myself, into one equal group of 4. What am I ?

   (A) 0   (B) 1

   (C) 3   (D) 4

2.   I am a number, I divide myself, into four equal groups of 1. What am I ?

   (A) 1   (B) 4

   (C) 6   (D) 2

3.   I am a number, I divide myself, into four equal groups of 2. What am I ?

   (A) 0   (B) 8

   (C) 2   (D) 4

4.   I am a number, I divide myself, into four equal groups of 3. What am I ?

   (A) 8   (B) 4

   (C) 12   (D) 3

5.   I am a number, I divide myself, into four equal groups of 4. What am I ?

   (A) 4   (B) 16

   (C) 6   (D) 2

www.math-knots.com

6. I am a number, I divide myself, into four equal groups of 5. What am I ?

(A) 4   (B) 8

(C) 20   (D) 16

7. I am a number, I divide myself, into four equal groups of 6. What am I ?

(A) 24   (B) 6

(C) 4   (D) 12

8. I am a number, I divide myself, into four equal groups of 7. What am I ?

(A) 7   (B) 4

(C) 16   (D) 28

9. I am a number, I divide myself, into four equal groups of 8. What am I ?

(A) 18   (B) 8

(C) 4   (D) 32

10. I am a number, I divide myself, into four equal groups of 9. What am I ?

(A) 36   (B) 4

(C) 9   (D) 32

11. I am a number, I divide myself, into four equal groups of 10. What am I ?

    (A) 28   (B) 40

    (C) 10   (D) 36

12. I am a number, I divide myself, into four equal groups of 11. What am I ?

    (A) 44   (B) 48

    (C) 40   (D) 11

13. I am a number, I divide myself, into four equal groups of 12. What am I ?

    (A) 12   (B) 36

    (C) 48   (D) 44

14. I am a number, I divide myself, into four equal groups of 13. What am I ?

    (A) 36   (B) 13

    (C) 52   (D) 48

15. I am a number, I divide myself, into four equal groups of 14. What am I ?

    (A) 36   (B) 52

    (C) 14   (D) 56

www.math-knots.com

Exercise - 4

Solve the maze run below.

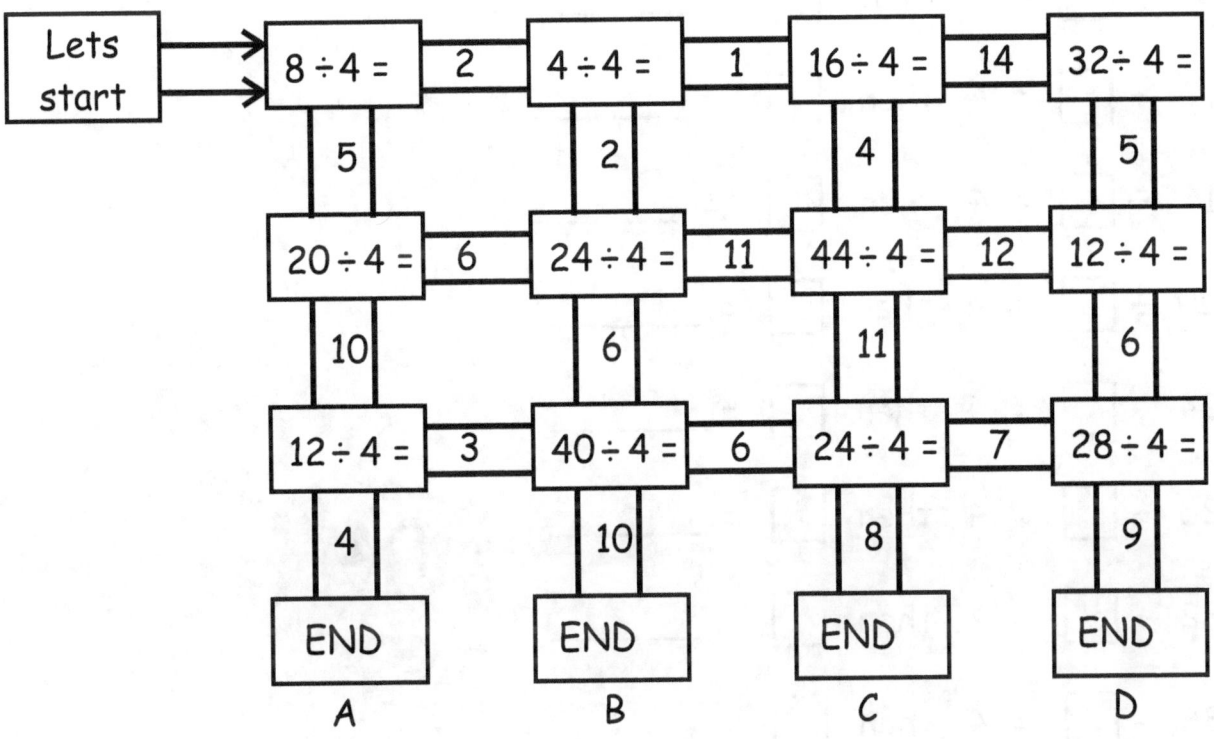

| Lets start | → → | 8 ÷ 4 = | 2 | 4 ÷ 4 = | 1 | 16 ÷ 4 = | 14 | 32 ÷ 4 = |

Who won the race ? _____

www.math-knots.com

Exercise - 5

1. 4 ÷ ☐ = 1   then   ☐ = _____

2. 8 ÷ ☐ = 4   then   ☐ = _____

3. 12 ÷ ☐ = 4   then   ☐ = _____

4. 16 ÷ ☐ = 4   then   ☐ = _____

5. 20 ÷ ☐ = 4   then   ☐ = _____

6. 24 ÷ ☐ = 4   then   ☐ = _____

7. 28 ÷ ☐ = 4   then   ☐ = _____

8. 32 ÷ ☐ = 4   then   ☐ = _____

9. 36 ÷ ☐ = 4   then   ☐ = _____

10. 40 ÷ ☐ = 4   then   ☐ = _____

11. 44 ÷ ☐ = 4   then   ☐ = _____

12. 48 ÷ ☐ = 4   then   ☐ = _____

Hey you are an expert of division facts of #4 !!!

www.math-knots.com

Division is opposite of Multiplication.
Division is splitting into equal parts or groups or equal sharing or equal partitioning.
Dividend: The dividend is the number that is being divided in the division process.
Divisor: The number by which dividend is being divided by is called divisor.
Quotient: A quotient is a result obtained in division process.

$$10 \div 5 = 2$$

Dividend. Divisor. Quotient
Let's learn division facts for #5

www.math-knots.com

1. Lets learn 5 ÷ 1 = 5

A.

B.
$$\div \quad 5$$

1 |

=

C.
┌──────────────┐
│ **5 ÷ 1 = 5** │
└──────────────┘

2. Lets learn 10 ÷ 5 = 2

A.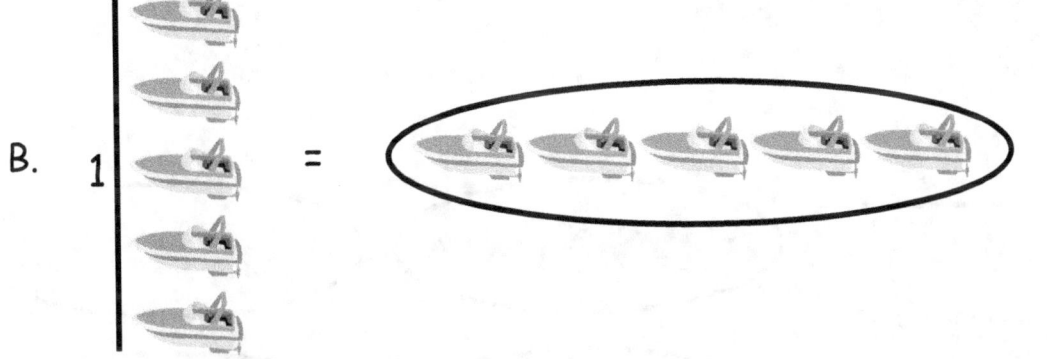

B.
$$\div \quad 2$$

5 |

=

C.
┌──────────────┐
│ **10 ÷ 5 = 2** │
└──────────────┘

www.math-knots.com

3.   Lets learn 15 ÷ 5 = 3

A.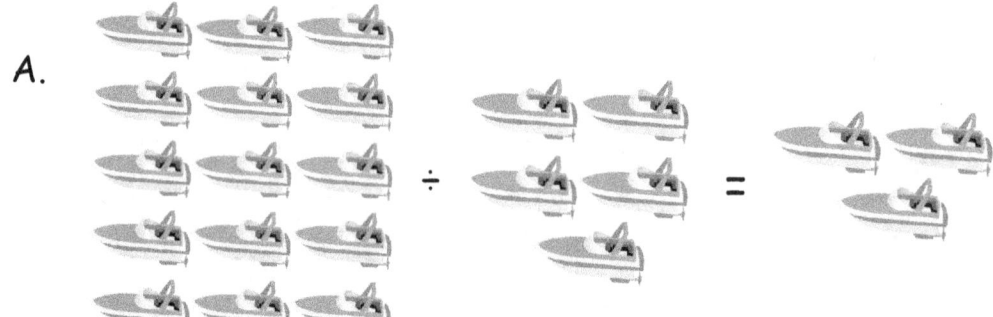

B.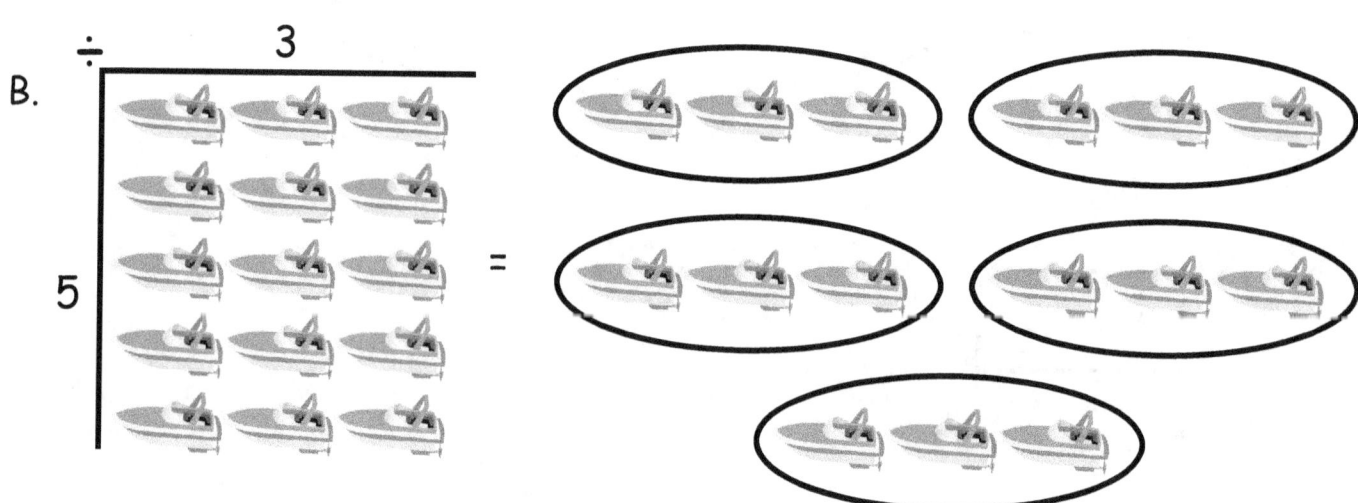

C.   15 ÷ 5 = 3

Did you know you can write division sign
in three different ways
÷ , / and —

www.math-knots.com

4. Lets learn $20 \div 5 = 4$

A.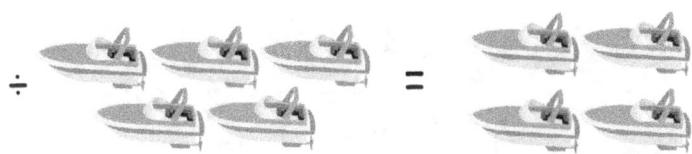

B.

$\div$ 4

5

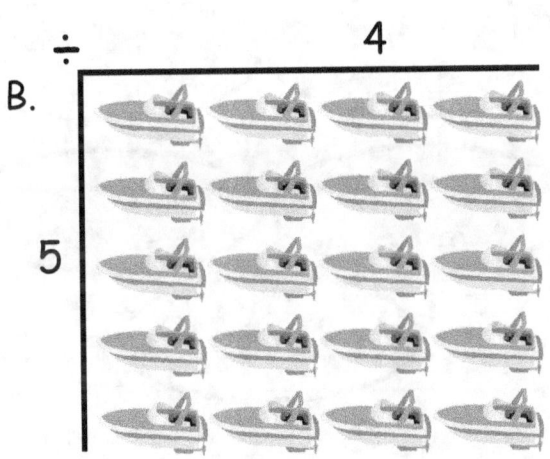

=

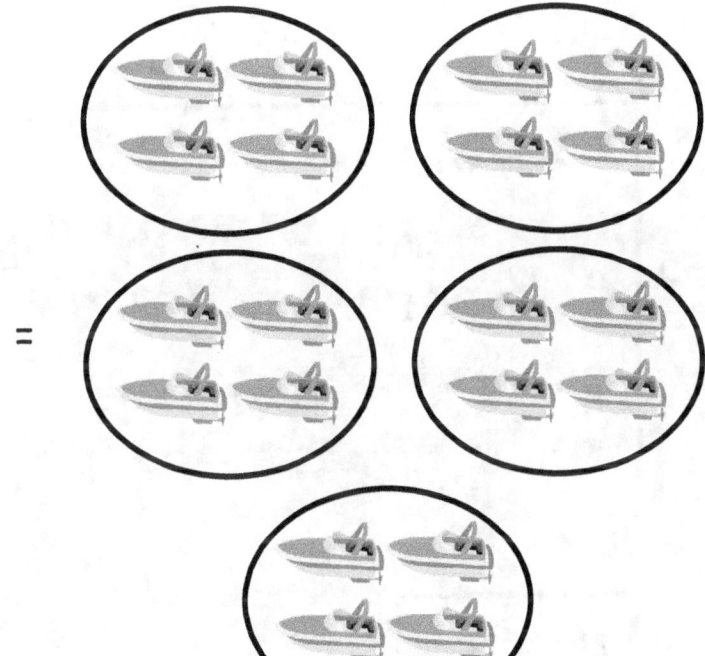

C. $\boxed{20 \div 5 = 4}$

Did You Know...?

Did you know division by 5 means
Dividing the given number into 5 equal share's ?
Dividing the number into five equal Groups.

www.math-knots.com

5. Lets learn 25 ÷ 5 = 5

A.

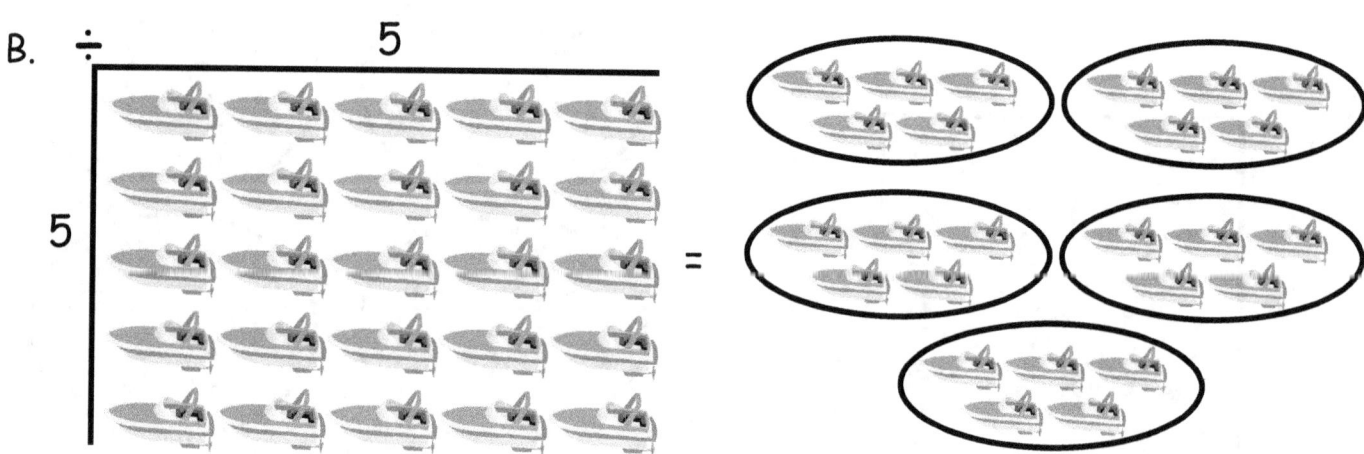

B. ÷

5

5

=

C.

| 25 ÷ 5 = 5 |

Did you know division is splitting a number up by any give number.

www.math-knots.com

6.　Lets learn $30 \div 5 = 6$

A.

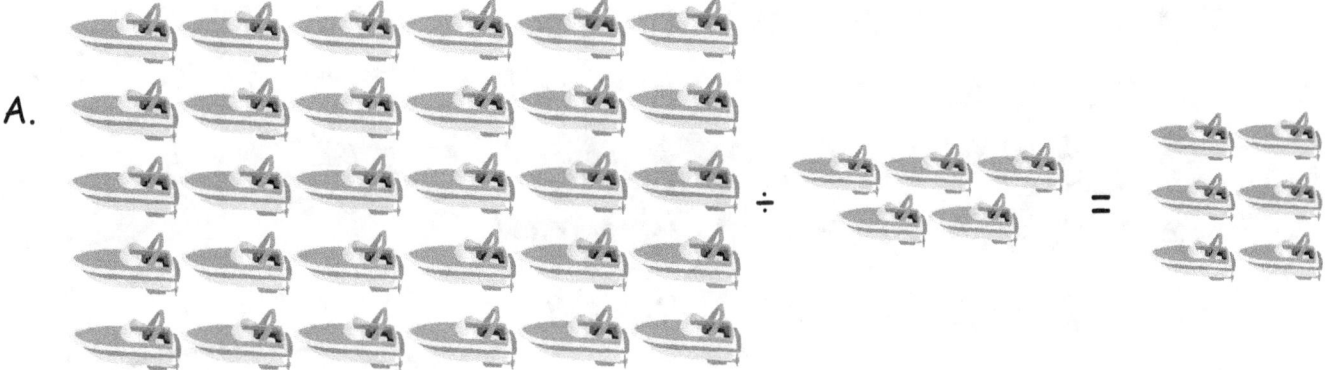

B.

$\div$
6
5

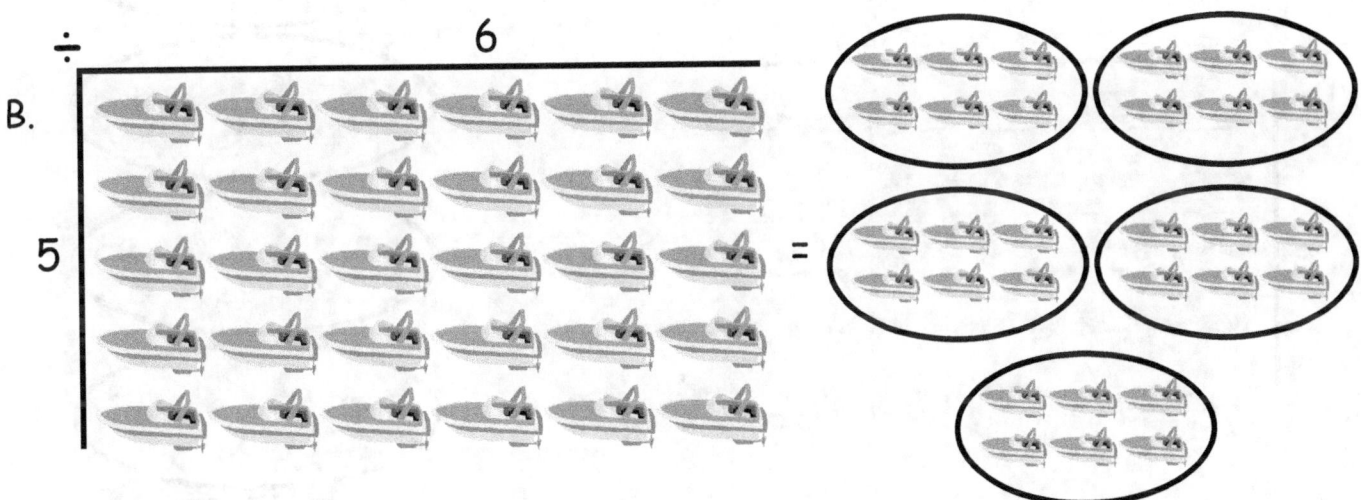

C.

$$30 \div 5 = 6$$

7. Lets learn 35 ÷ 5 = 7

A.

B.

÷

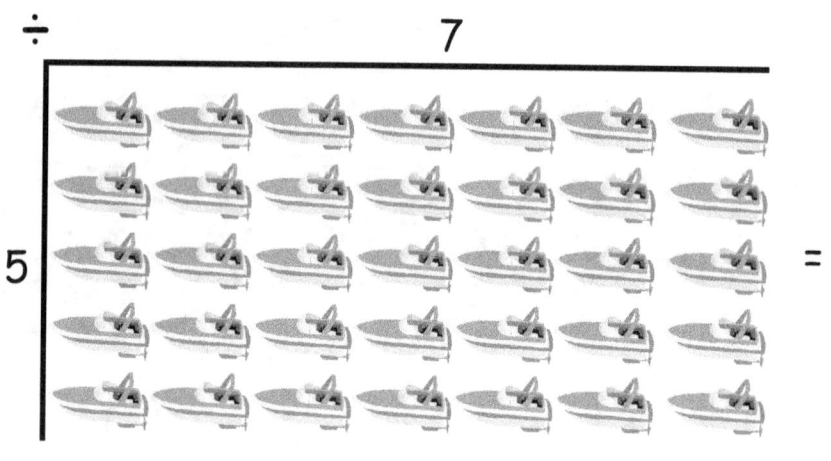

5

=

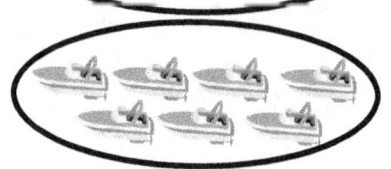
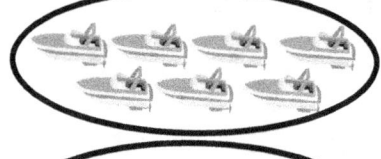

C. | 35 ÷ 5 = 7 |

www.math-knots.com

8.   Lets learn $40 \div 5 = 8$

A.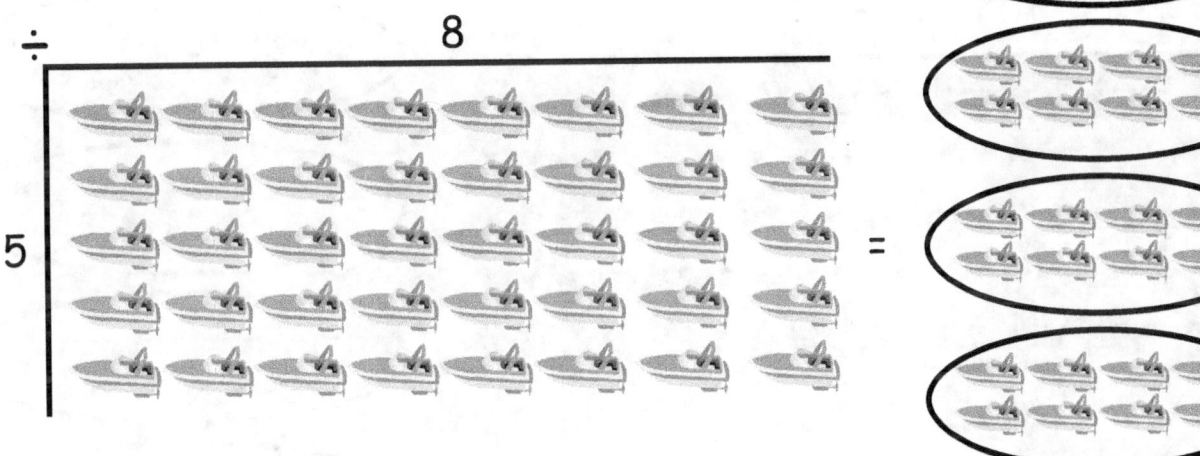

B.

C. | $40 \div 5 = 8$ |

9.   Lets learn $45 \div 5 = 9$

A.

B.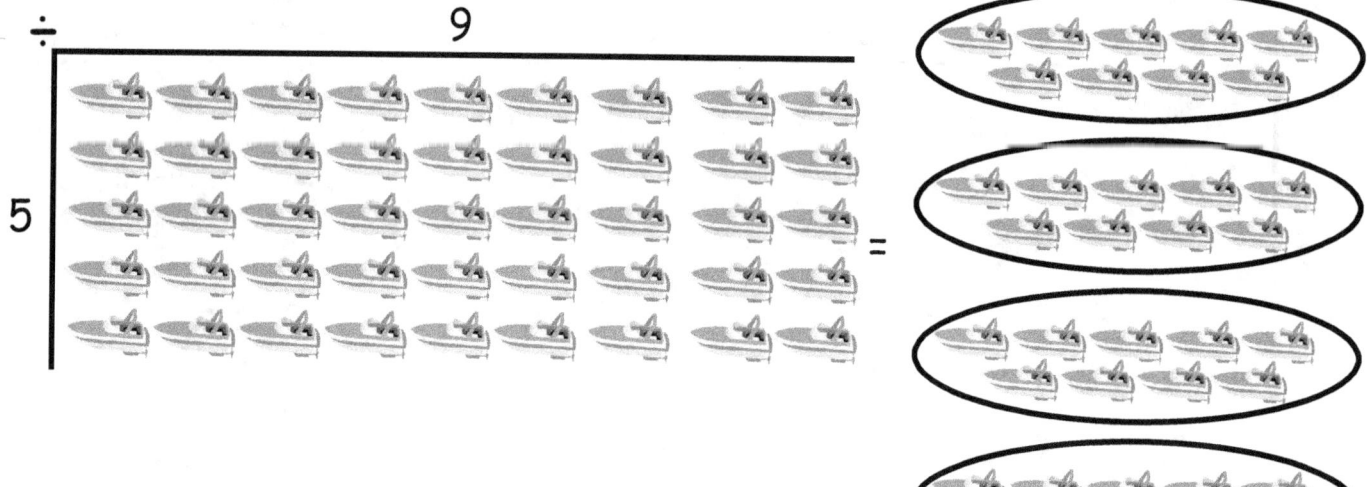

C. $\boxed{45 \div 5 = 9}$

10.   Lets learn $50 \div 5 = 5$

A.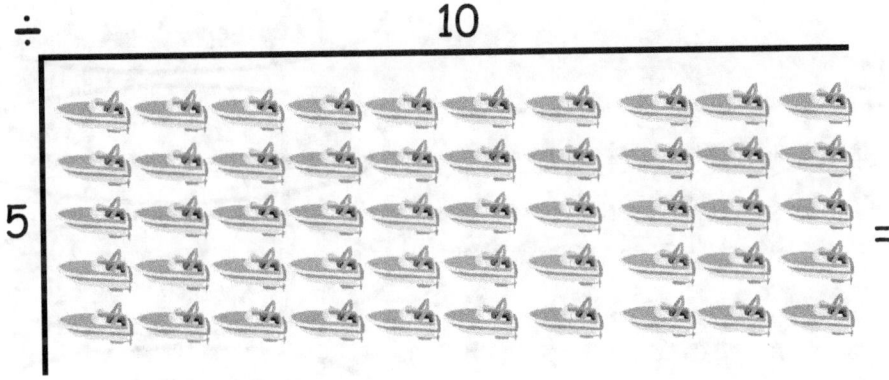

B.

$\div$

10

5

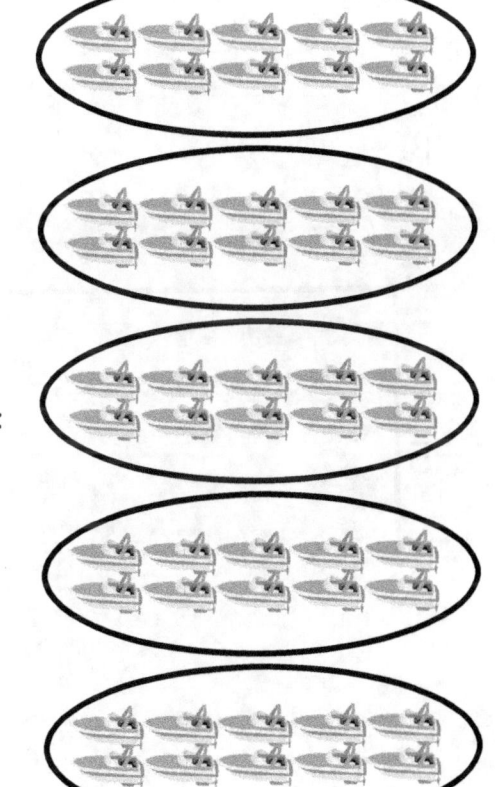

C. $\boxed{50 \div 5 = 10}$

11.   Lets learn $55 \div 5 = 11$

A.

B.

$\div$

11

5

$=$

C. | $55 \div 5 = 11$ |

12.   Lets learn 60 ÷ 5 = 12

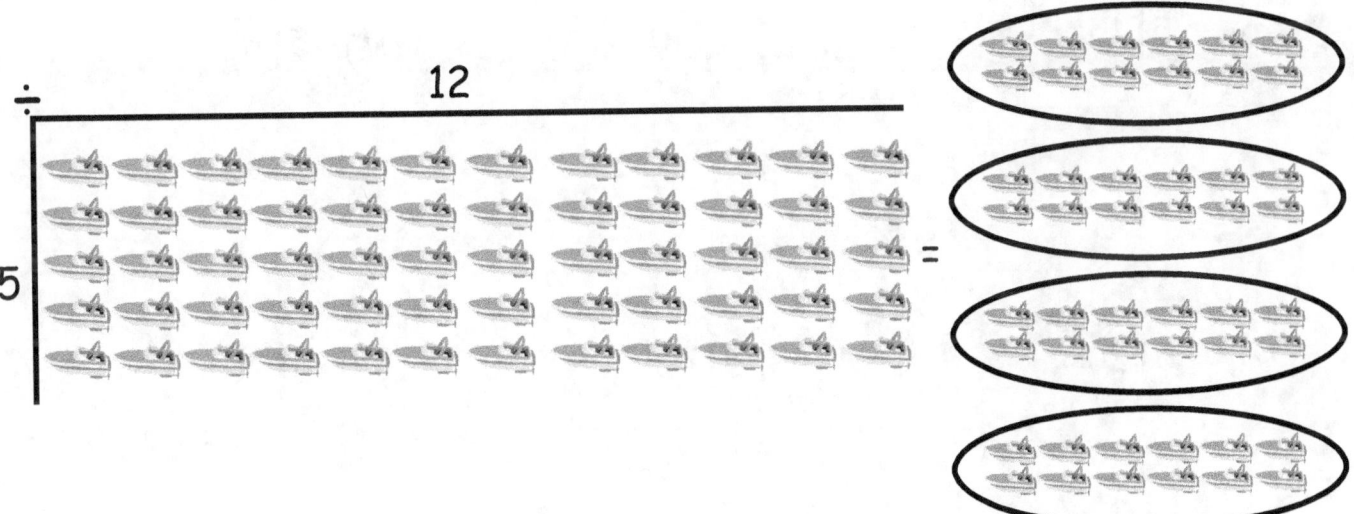

A.

B.

12

5

C.

$$60 \div 5 = 12$$

Exercise - 1

(A) $5\overline{)5}$

(B) $5\overline{)10}$

(C) $5\overline{)15}$

(D) $5\overline{)20}$

(E) $5\overline{)25}$

(F) $5\overline{)30}$

(G) $5\overline{)35}$

(H) $5\overline{)40}$

(I) $5\overline{)45}$

(J) $5\overline{)50}$

(K) $5\overline{)55}$

(L) $5\overline{)60}$

(M) $5\overline{)65}$

(N) $5\overline{)70}$

(O) $5\overline{)75}$

www.math-knots.com

Exercise - 2

1.	5 ÷ 5 =	_____
2.	10 ÷ 5 =	_____
3.	15 ÷ 5 =	_____
4.	20 ÷ 5 =	_____
5.	25 ÷ 5 =	_____
6.	30 ÷ 5 =	_____
7.	35 ÷ 5 =	_____
8.	40 ÷ 5 =	_____
9.	45 ÷ 5 =	_____
10.	50 ÷ 5 =	_____
11.	55 ÷ 5 =	_____
12.	60 ÷ 5 =	_____

1	× ____ =	5
2	× ____ =	10
3	× ____ =	15
4	× ____ =	20
5	× ____ =	25
6	× ____ =	30
7	× ____ =	35
8	× ____ =	40
9	× ____ =	45
10	× ____ =	50
11	× ____ =	55
12	× ____ =	60

Did you know division is splitting a
number up by any give number.

www.math-knots.com

Exercise - 3

1.   I am a number, I divide myself, into one equal group of 5. What am I ?

   (A) 0   (B) 1

   (C) 10   (D) 5

2.   I am a number, I divide myself, into five equal groups of 1. What am I ?

   (A) 10   (B) 5

   (C) 6   (D) 1

3.   I am a number, I divide myself, into five equal groups of 2. What am I ?

   (A) 5   (B) 10

   (C) 2   (D) 15

4.   I am a number, I divide myself, into five equal groups of 3. What am I ?

   (A) 5   (B) 3

   (C) 15   (D) 10

5.   I am a number, I divide myself, into five equal groups of 4. What am I ?

   (A) 20   (B) 15

   (C) 5   (D) 4

6.   I am a number, I divide myself, into five equal groups of 5. What am I ?

    (A)   15   (B)   10

    (C)   25   (D)   5

7.   I am a number, I divide myself, into five equal groups of 6. What am I ?

    (A)   30   (B)   6

    (C)   5   (D)   25

8.   I am a number, I divide myself, into five equal groups of 7. What am I ?

    (A)   30   (B)   7

    (C)   5   (D)   35

9.   I am a number, I divide myself, into five equal groups of 8. What am I ?

    (A)   20   (B)   5

    (C)   8   (D)   40

10.   I am a number, I divide myself, into five equal groups of 9. What am I ?

    (A)   15   (B)   9

    (C)   45   (D)   5

11.   I am a number, I divide myself, into five equal groups of 10. What am I ?

   (A) 5   (B) 50

   (C) 25   (D) 10

12.   I am a number, I divide myself, into five equal groups of 11. What am I ?

   (A) 55   (B) 33

   (C) 40   (D) 11

13.   I am a number, I divide myself, into five equal groups of 12. What am I ?

   (A) 5   (B) 12

   (C) 48   (D) 60

14.   I am a number, I divide myself, into five equal groups of 13. What am I ?

   (A) 45   (B) 5

   (C) 65   (D) 13

15.   I am a number, I divide myself, into five equal groups of 14. What am I ?

   (A) 35   (B) 5

   (C) 14   (D) 70

Exercise - 4

Solve the maze run below.

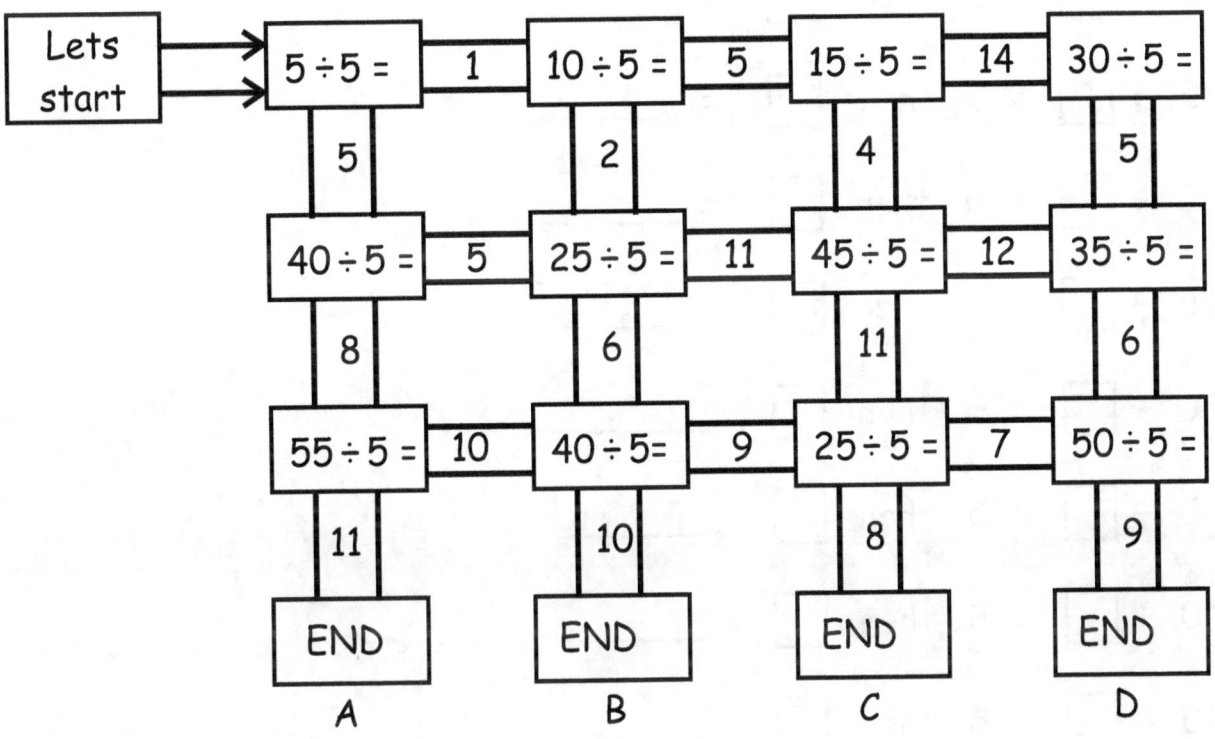

Lets start	5 ÷ 5 =	1	10 ÷ 5 =	5	15 ÷ 5 =	14	30 ÷ 5 =

5 · 2 · 4 · 5

40 ÷ 5 = · 5 · 25 ÷ 5 = · 11 · 45 ÷ 5 = · 12 · 35 ÷ 5 =

8 · 6 · 11 · 6

55 ÷ 5 = · 10 · 40 ÷ 5= · 9 · 25 ÷ 5 = · 7 · 50 ÷ 5 =

11 · 10 · 8 · 9

END A · END B · END C · END D

Who won the race ? _____

www.math-knots.com

Exercise - 5

1. $5 \div \square = 1$ then $\square = $ _____

2. $10 \div \square = 5$ then $\square = $ _____

3. $15 \div \square = 5$ then $\square = $ _____

4. $20 \div \square = 5$ then $\square = $ _____

5. $25 \div \square = 5$ then $\square = $ _____

6. $30 \div \square = 5$ then $\square = $ _____

7. $35 \div \square = 5$ then $\square = $ _____

8. $40 \div \square = 5$ then $\square = $ _____

9. $45 \div \square = 5$ then $\square = $ _____

10. $50 \div \square = 5$ then $\square = $ _____

11. $55 \div \square = 5$ then $\square = $ _____

12. $60 \div \square = 5$ then $\square = $ _____

Hey you are an expert of division facts of #5 !!!

www.math-knots.com

Division is opposite of Multiplication.
Division is splitting into equal parts or groups or equal sharing or equal partitioning.
Dividend: The dividend is the number that is being divided in the division process.
Divisor: The number by which dividend is being divided by is called divisor.
Quotient: A quotient is a result obtained in division process.

$$12 \div 6 = 2$$

Dividend. Divisor. Quotient
Let's learn division facts for #6

www.math-knots.com

www.math-knots.com

1.   Lets learn 6 ÷ 1 = 6

A.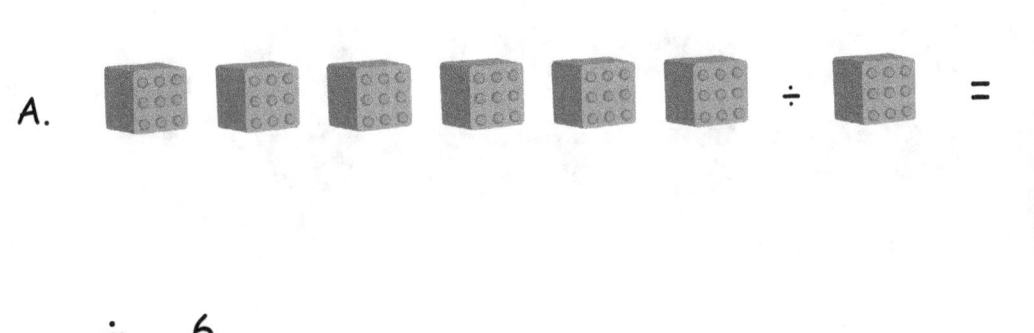

B.

C. | **6 ÷ 1 = 6** |

Did You Know...?

Did you know you can write division sign
in three different ways
÷ , / and —

www.math-knots.com

2. Lets learn $12 \div 6 = 2$

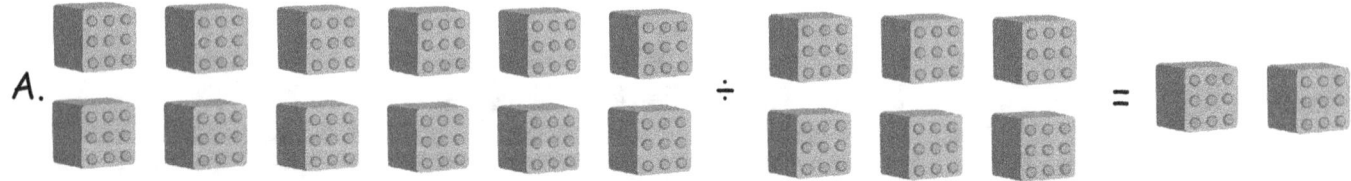

A.

B.

C. | $12 \div 6 = 2$ |

Did you know division by 6 means
Dividing the given number into 6 equal share's ?
Dividing the number into six equal Groups.

3.   Lets learn 18 ÷ 6 = 3

A.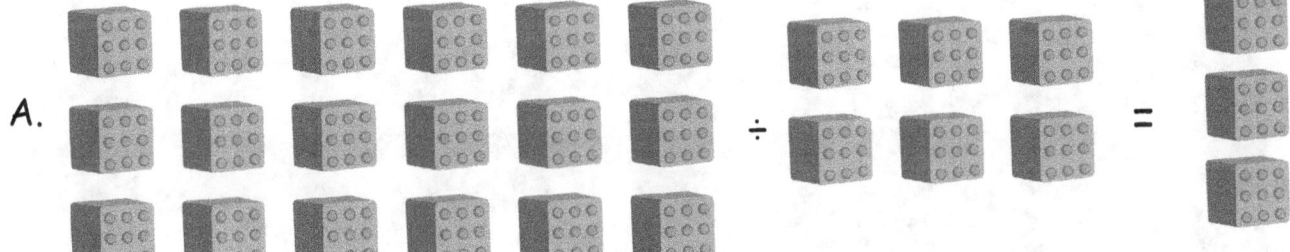

B.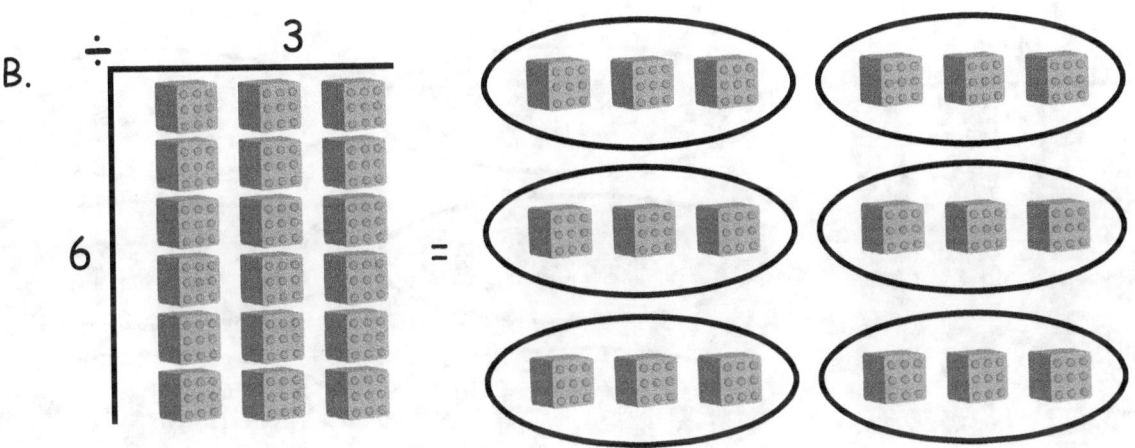

C. | 18 ÷ 6 = 3 |

Did you know division is splitting a number up by any give number.

www.math-knots.com

4.   Lets learn $24 \div 6 = 4$

A.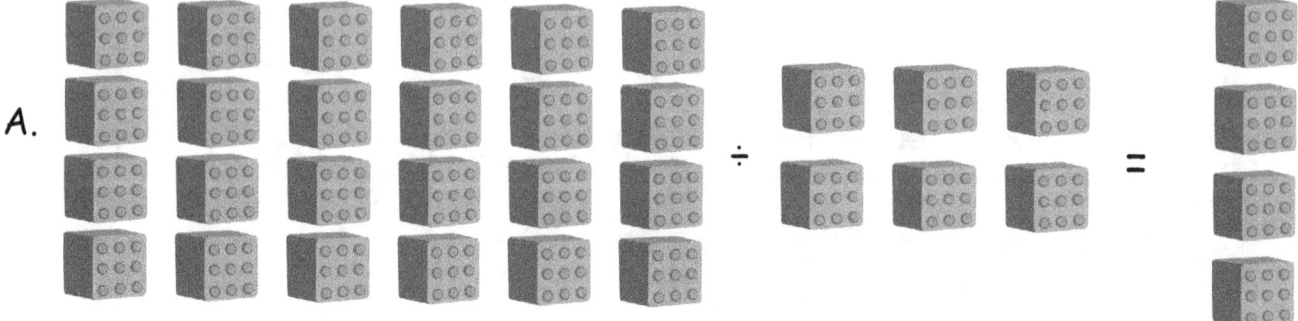

B.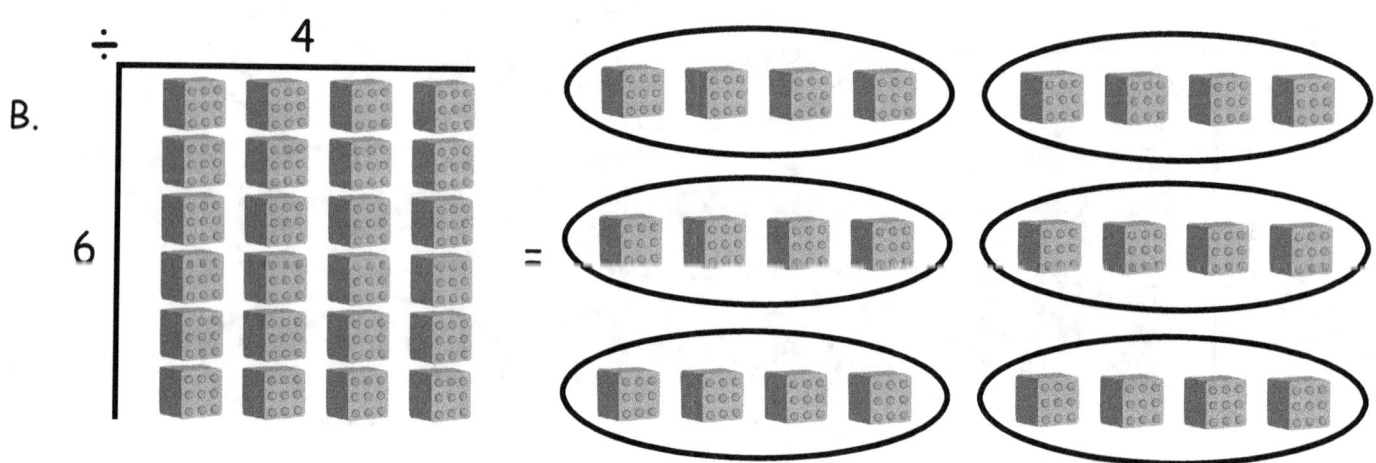

C. $\boxed{24 \div 6 = 4}$

Did you know division by 6 means
Dividing the given number into 6 equal half's ?
Dividing the number into equal Groups of six's.

www.math-knots.com

5.   Lets learn 30 ÷ 6 = 5

A.

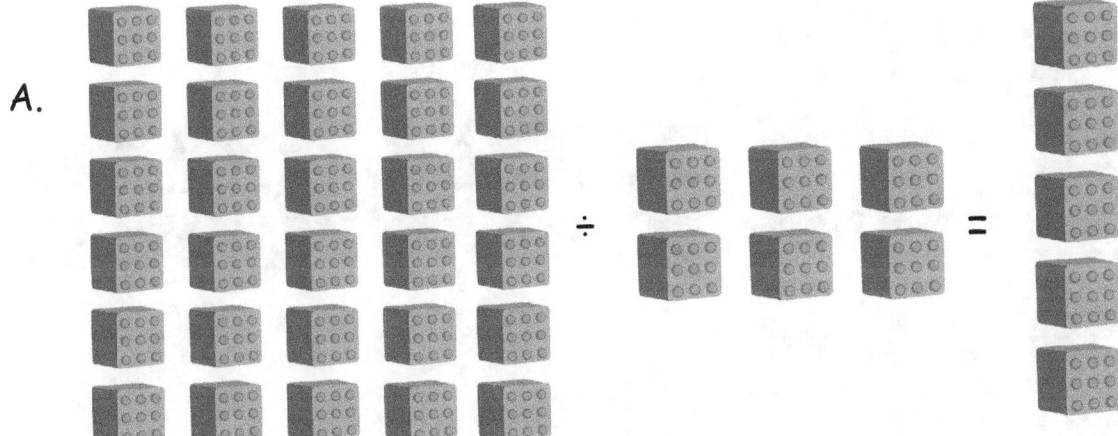

B.

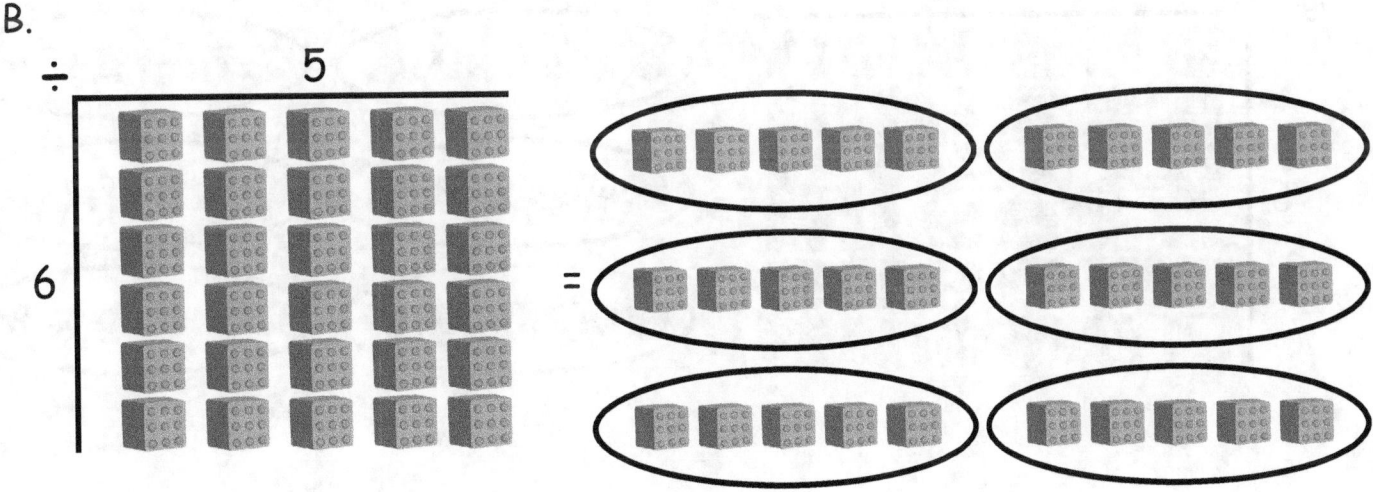

C. $30 \div 6 = 5$

6.   Lets learn 36 ÷ 6 = 6

A.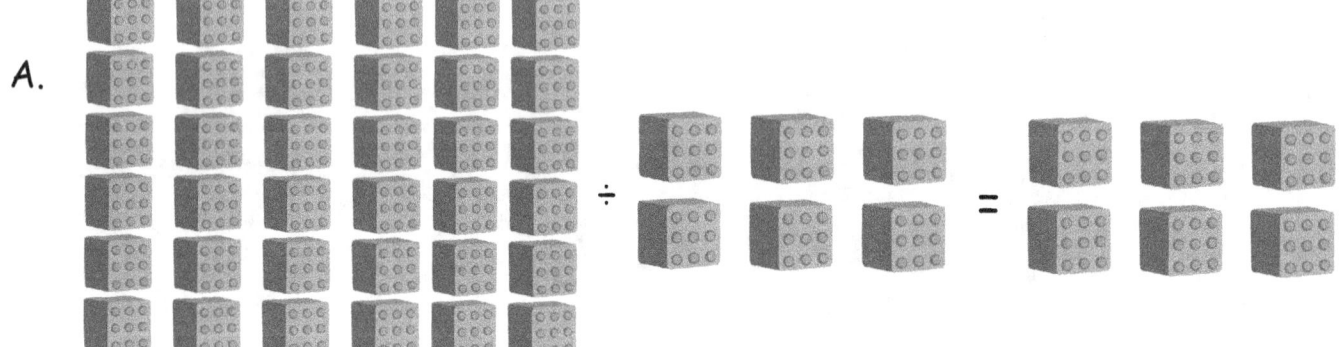

B.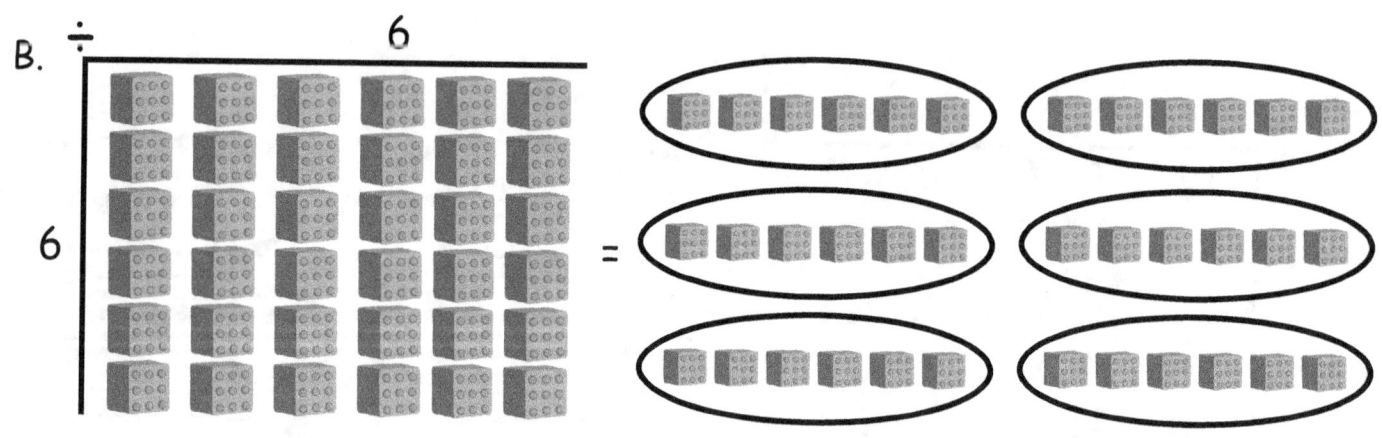

C. | **36 ÷ 6 = 6** |

7. Lets learn 42 ÷ 6 = 7

A.

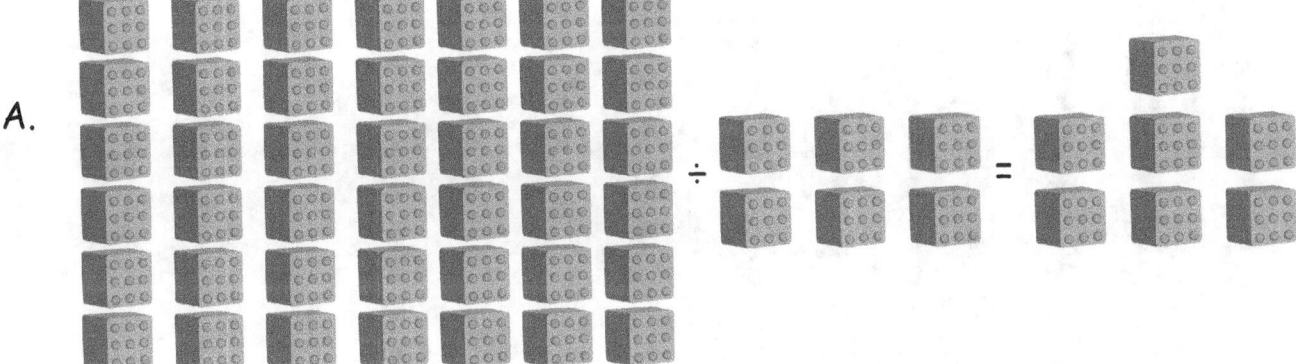

B.

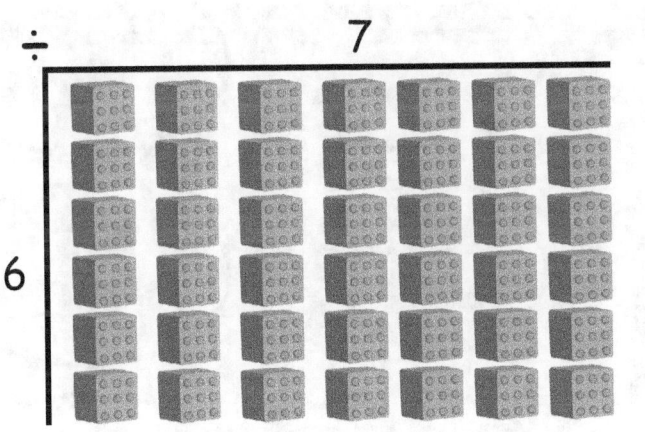

=

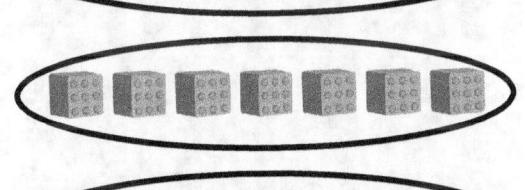

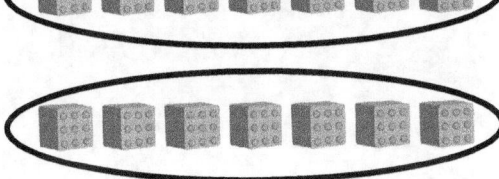

C.

$$42 \div 6 = 7$$

www.math-knots.com

8. Lets learn $48 \div 6 = 8$

A.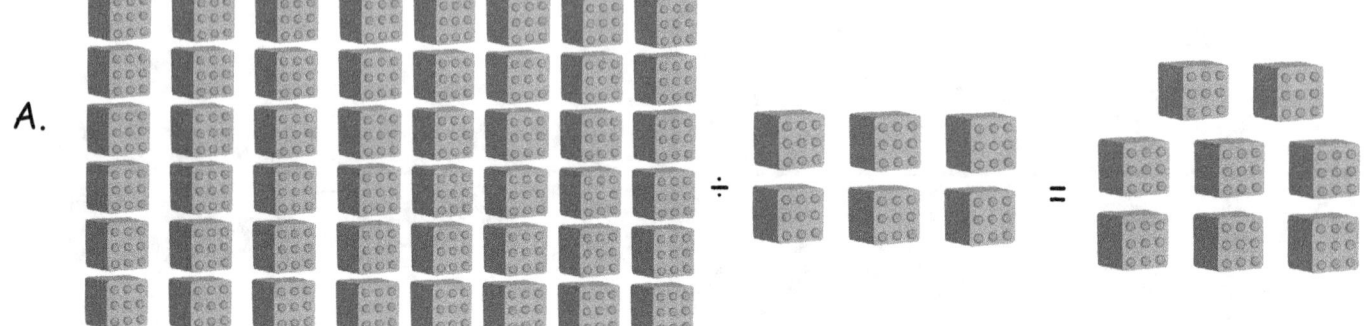

B.

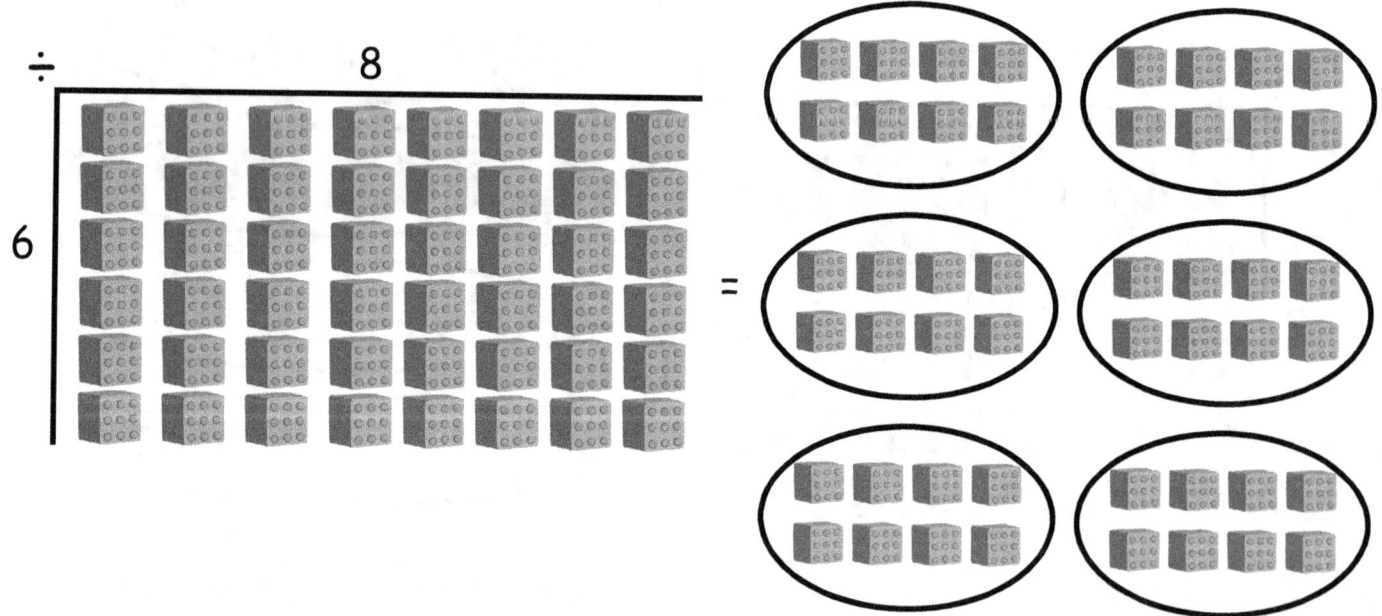

C. $\boxed{48 \div 6 = 8}$

9. Lets learn $54 \div 6 = 9$

A.

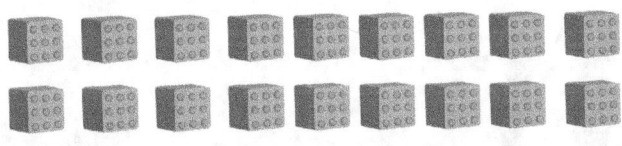

B.

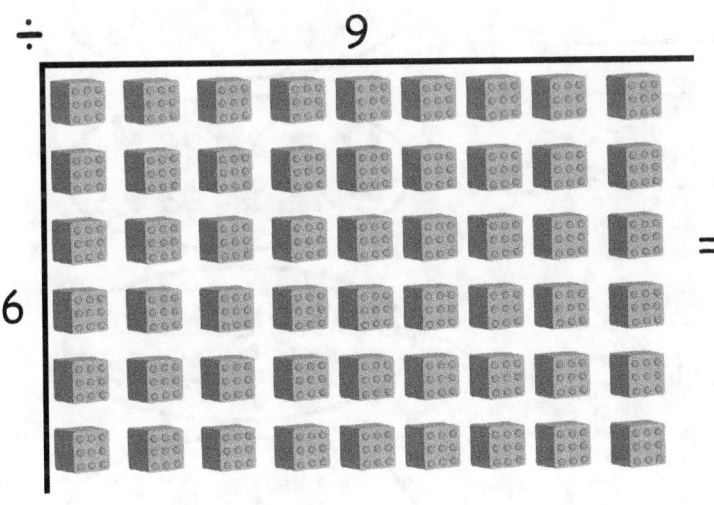

C. $\boxed{54 \div 6 = 9}$

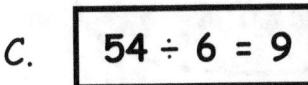

www.math-knots.com

10. Lets learn 60 ÷ 6 = 10

A.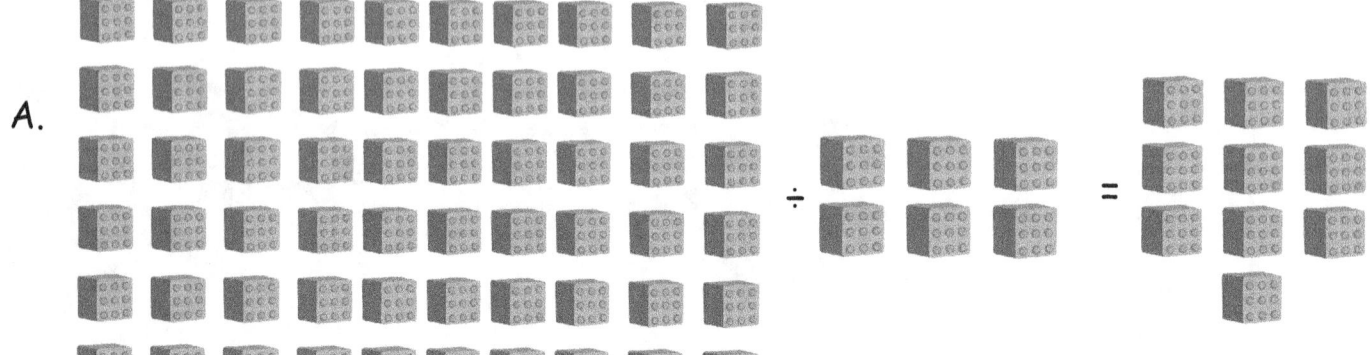

B.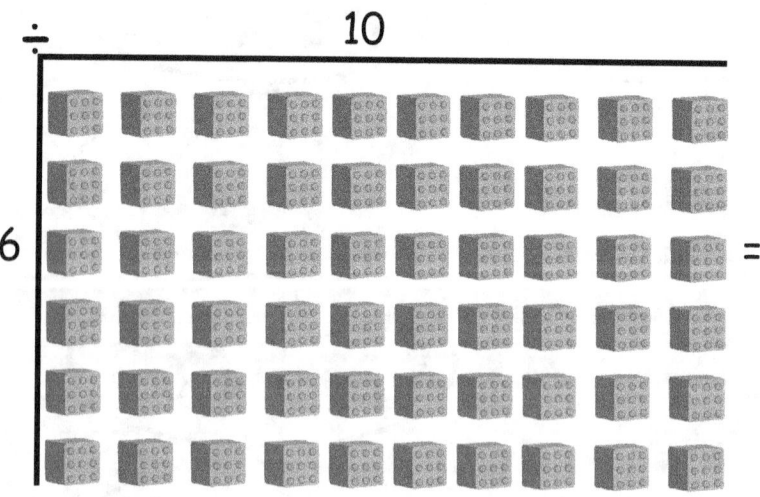

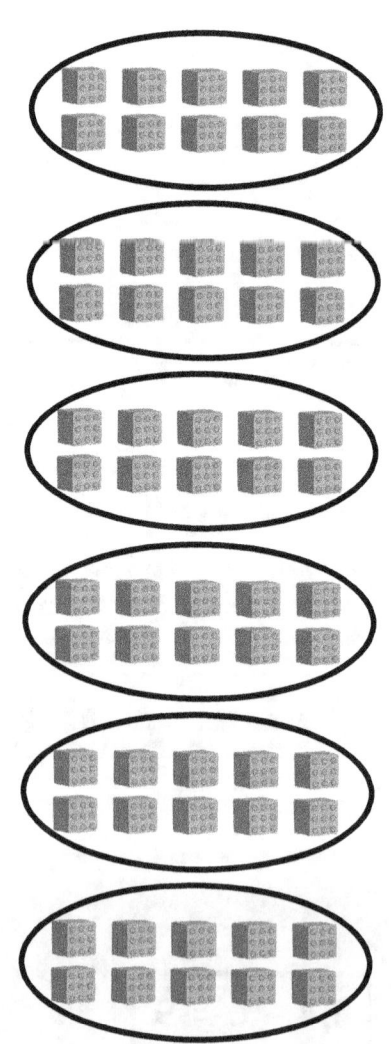

C. $60 \div 6 = 10$

11. Lets learn $66 \div 6 = 11$

A.

B.

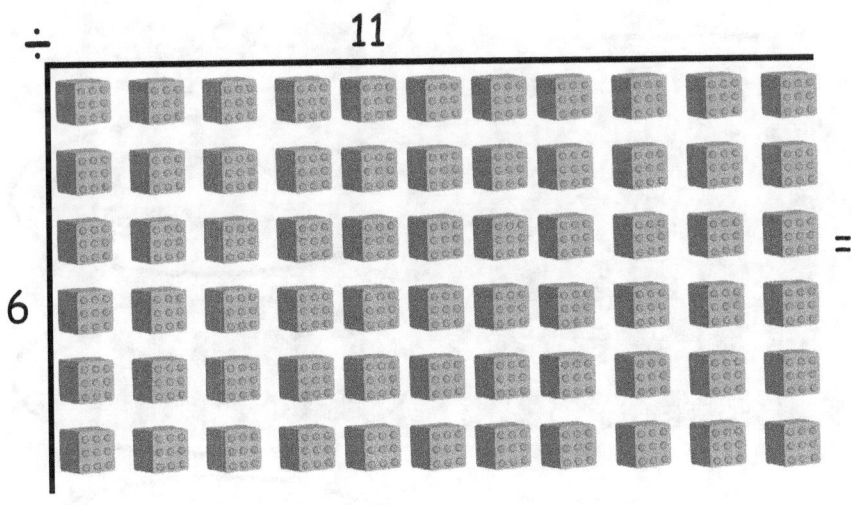

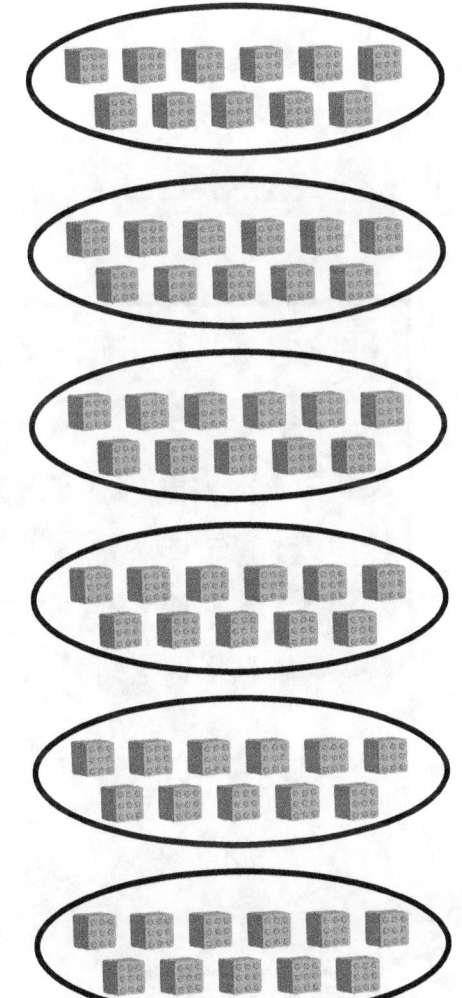

C. $\boxed{66 \div 6 = 11}$

DIVISION FACTS

12. Lets learn $72 \div 6 = 12$

A.

B.

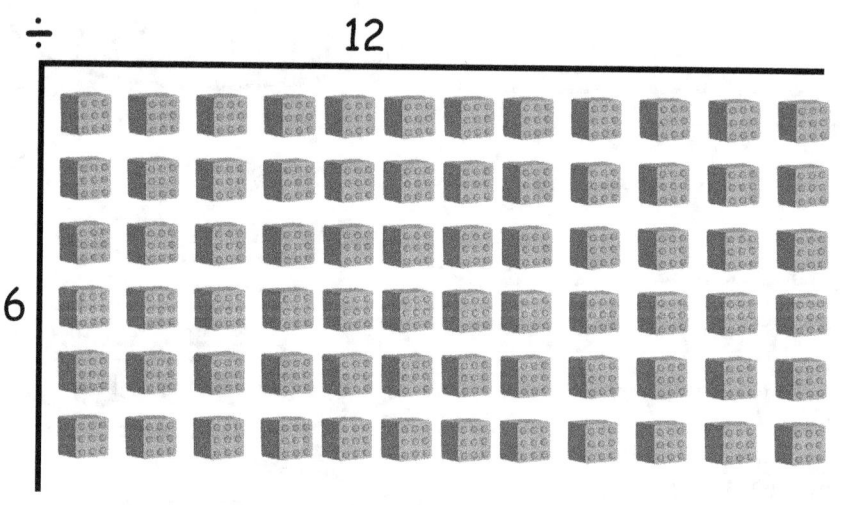

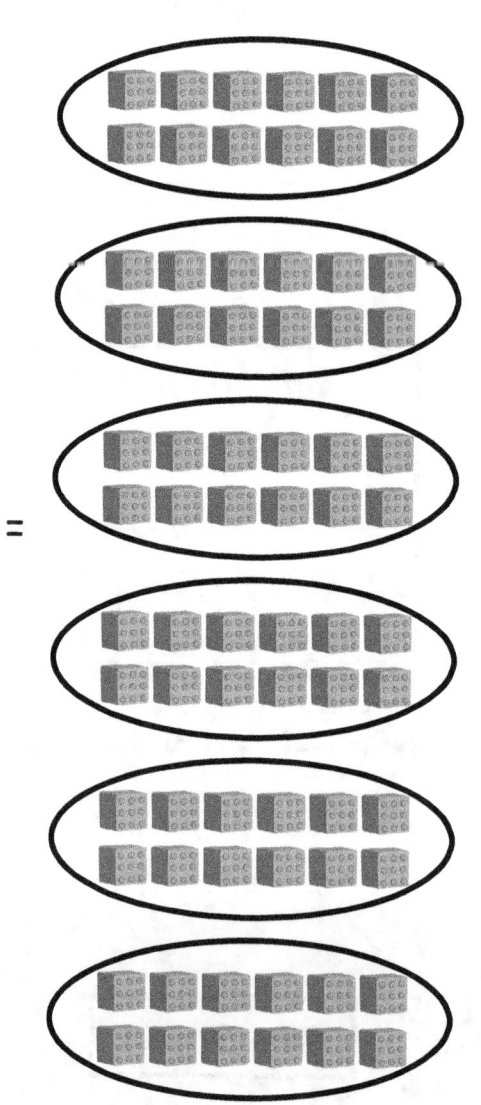

C. $\boxed{72 \div 6 = 12}$

www.math-knots.com

Exercise - 1

(A) $6\overline{)6}$

(F) $6\overline{)36}$

(K) $6\overline{)66}$

(B) $6\overline{)12}$

(G) $6\overline{)42}$

(L) $6\overline{)72}$

(C) $6\overline{)18}$

(H) $6\overline{)48}$

(M) $6\overline{)78}$

(D) $6\overline{)24}$

(I) $6\overline{)54}$

(N) $6\overline{)84}$

(E) $6\overline{)30}$

(J) $6\overline{)60}$

(O) $6\overline{)90}$

Exercise - 2

1.	6 ÷ 6	=	_____	
2.	12 ÷ 6	=	_____	
3.	18 ÷ 6	=	_____	
4.	24 ÷ 6	=	_____	
5.	30 ÷ 6	=	_____	
6.	36 ÷ 6	=	_____	
7.	42 ÷ 6	=	_____	
8.	48 ÷ 6	=	_____	
9.	54 ÷ 6	=	_____	
10.	60 ÷ 6	=	_____	
11.	66 ÷ 6	=	_____	
12.	72 ÷ 6	=	_____	

1	× ___	=	6
2	× ___	=	12
3	× ___	=	18
4	× ___	=	24
5	× ___	=	30
6	× ___	=	36
7	× ___	=	42
8	× ___	=	48
9	× ___	=	54
10	× ___	=	60
11	× ___	=	66
12	× ___	=	72

Did you know division is splitting a number up by any give number.

www.math-knots.com

 **Exercise - 3**

1. I am a number, I divide myself, into one equal group of 6. What am I ?

   (A) 0   (B) 1

   (C) 10   (D) 6

2. I am a number, I divide myself, into six equal groups of 1. What am I ?

   (A) 1   (B) 16

   (C) 6   (D) 12

3. I am a number, I divide myself, into six equal groups of 2. What am I ?

   (A) 6   (B) 12

   (C) 2   (D) 1

4. I am a number, I divide myself, into six equal groups of 3. What am I ?

   (A) 8   (B) 3

   (C) 18   (D) 6

5. I am a number, I divide myself, into six equal groups of 4. What am I ?

   (A) 24   (B) 14

   (C) 6   (D) 12

                   www.math-knots.com

6. I am a number, I divide myself, into six equal groups of 5. What am I ?

    (A) 12   (B) 6

    (C) 30   (D) 5

7. I am a number, I divide myself, into six equal groups of 6. What am I ?

    (A) 30   (B) 36

    (C) 6   (D) 12

8. I am a number, I divide myself, into six equal groups of 7. What am I ?

    (A) 42   (B) 24

    (C) 6   (D) 7

9. I am a number, I divide myself, into six equal groups of 8. What am I ?

    (A) 8   (B) 6

    (C) 30   (D) 48

10. I am a number, I divide myself, into six equal groups of 9. What am I ?

    (A) 36   (B) 6

    (C) 9   (D) 54

www.math-knots.com

11.   I am a number, I divide myself, into six equal groups of 10. What am I ?

(A) 6   (B) 60

(C) 10   (D) 54

12.   I am a number, I divide myself, into six equal groups of 11. What am I ?

(A) 66   (B) 48

(C) 11   (D) 6

13.   I am a number, I divide myself, into six equal groups of 12. What am I ?

(A) 32   (B) 36

(C) 72   (D) 60

14.   I am a number, I divide myself, into six equal groups of 13. What am I ?

(A) 36   (B) 6

(C) 13   (D) 78

15.   I am a number, I divide myself, into six equal groups of 14. What am I ?

(A) 14   (B) 52

(C) 84   (D) 6

101

Exercise - 4

Solve the maze run below.

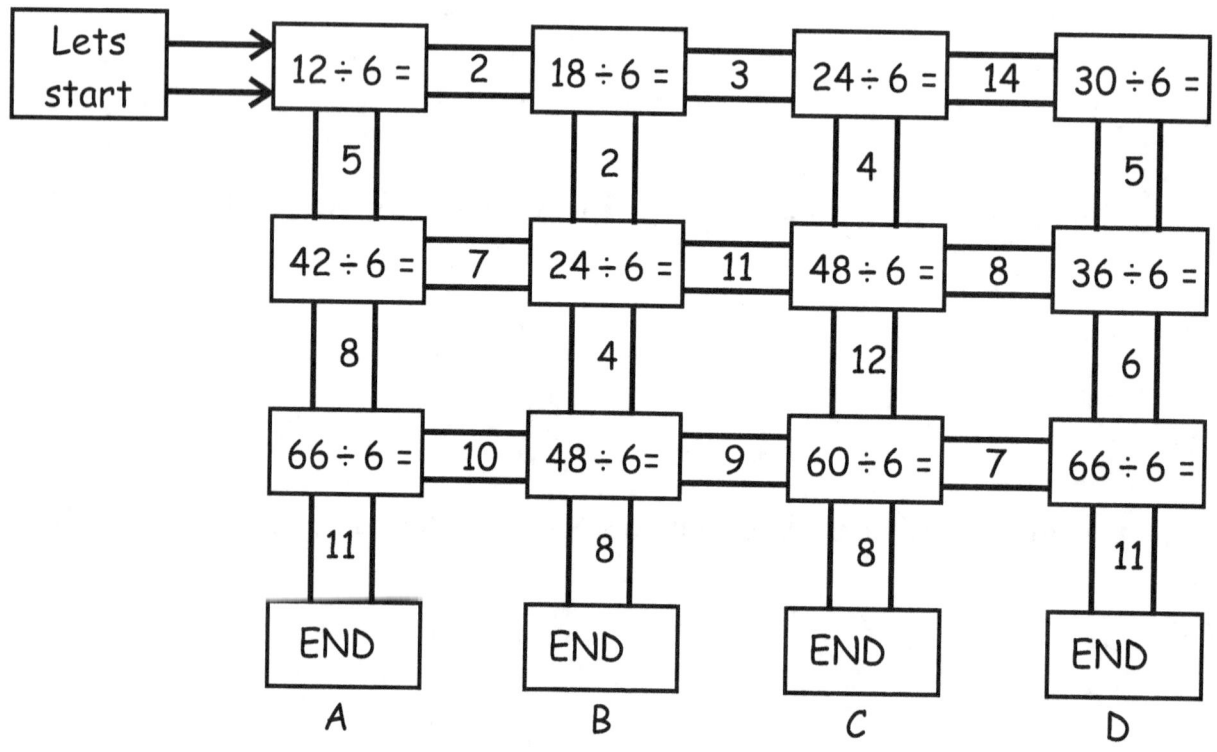

	A	B	C	D
Lets start →	12 ÷ 6 = 2	18 ÷ 6 = 3	24 ÷ 6 = 14	30 ÷ 6 =
	5	2	4	5
	42 ÷ 6 = 7	24 ÷ 6 = 11	48 ÷ 6 = 8	36 ÷ 6 =
	8	4	12	6
	66 ÷ 6 = 10	48 ÷ 6 = 9	60 ÷ 6 = 7	66 ÷ 6 =
	11	8	8	11
	END A	END B	END C	END D

Who won the race ? _____

www.math-knots.com

Exercise - 5

1. 6 ÷ ☐ = 1 then ☐ = _____

2. 12 ÷ ☐ = 6 then ☐ = _____

3. 18 ÷ ☐ = 6 then ☐ = _____

4. 24 ÷ ☐ = 6 then ☐ = _____

5. 30 ÷ ☐ = 6 then ☐ = _____

6. 36 ÷ ☐ = 6 then ☐ = _____

7. 42 ÷ ☐ = 6 then ☐ = _____

8. 48 ÷ ☐ = 6 then ☐ = _____

9. 54 ÷ ☐ = 6 then ☐ = _____

10. 60 ÷ ☐ = 6 then ☐ = _____

11. 66 ÷ ☐ = 6 then ☐ = _____

12. 72 ÷ ☐ = 6 then ☐ = _____

Hey you are an expert of division facts of #6 !!!

www.math-knots.com

Division is opposite of Multiplication.
Division is splitting into equal parts or groups or
equal sharing or equal partitioning.
Dividend: The dividend is the number that is being
divided in the division process.
Divisor: The number by which dividend is being
divided by is called divisor.
Quotient: A quotient is a result obtained in
division process.

$$14 \div 7 = 2$$

Dividend. Divisor. Quotient
Let's learn division facts for #7

www.math-knots.com

1. Lets learn 7 ÷ 1 = 7

A.  ÷ 🦆 =

B.

C. | 7 ÷ 1 = 7 |

Did you know you can write division sign
in three different ways
÷ , / and —

www.math-knots.com

2.   Lets learn 14 ÷ 7 = 2

A.

B.

$$\overset{\div \quad 2}{\underline{7 \, \Big|}}$$

C. | 14 ÷ 7 = 2 |

Did you know division by 7 means
Dividing the given number into 7 equal share's ?
Dividing the number into seven equal Groups.

www.math-knots.com

3. Lets learn 21 ÷ 7 = 3

A.

B.

C.

21 ÷ 7 = 3

Did you know division is splitting a
number up by any give number.

www.math-knots.com

4.   Lets learn 28 ÷ 7 = 4

A.

B.

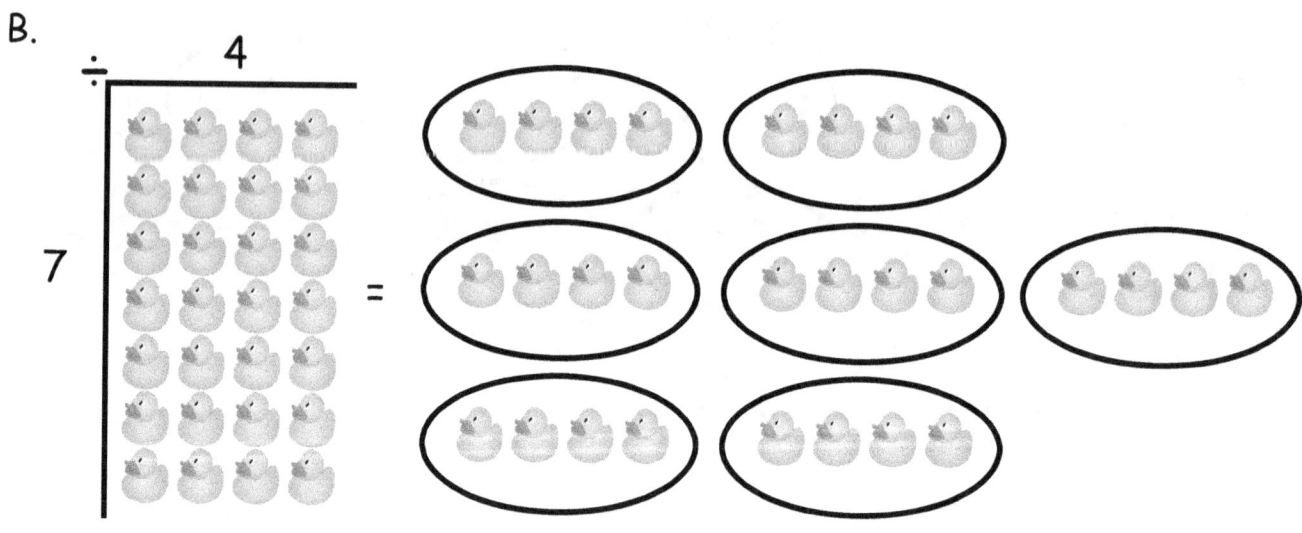

C.   **28 ÷ 7 = 4**

5. Lets learn 35 ÷ 7 = 5

A.

B.

C.

35 ÷ 7 = 5

111   www.math-knots.com

6. Lets learn $42 \div 7 = 6$

A.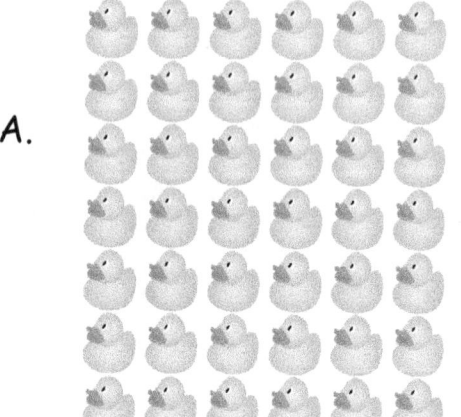

B.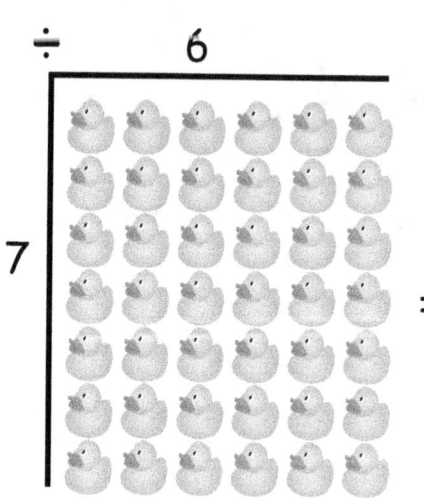

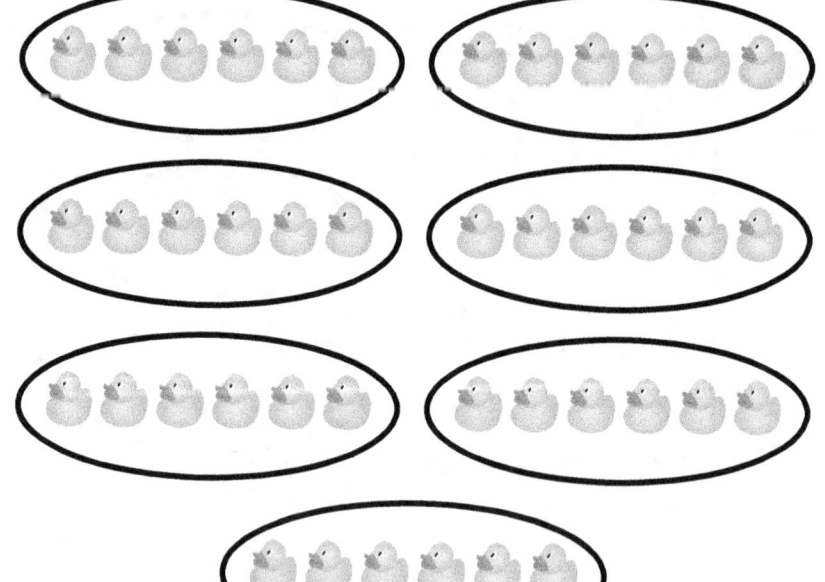

C. $\boxed{42 \div 7 = 6}$

www.math-knots.com

7.   Lets learn 49 ÷ 7 = 7

A.
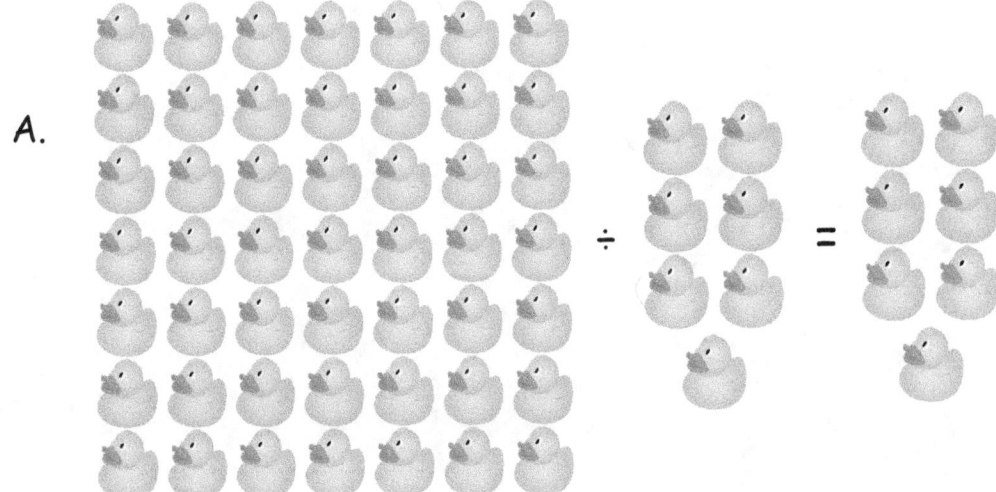

B.
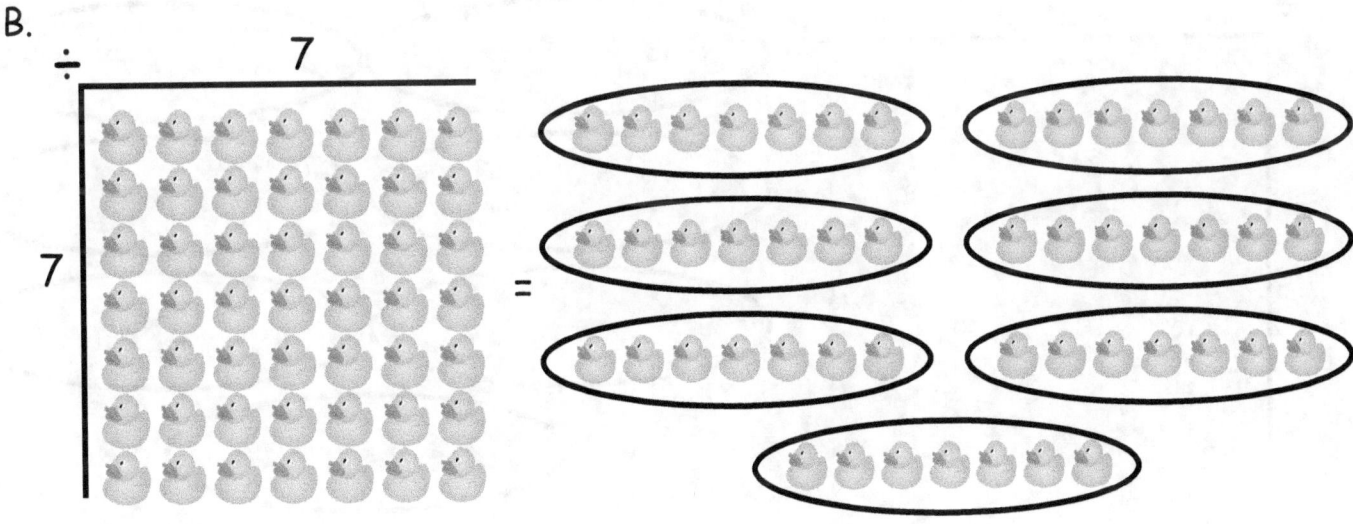

C.   **49 ÷ 7 = 7**

8. Lets learn 56 ÷ 7 = 8

A.

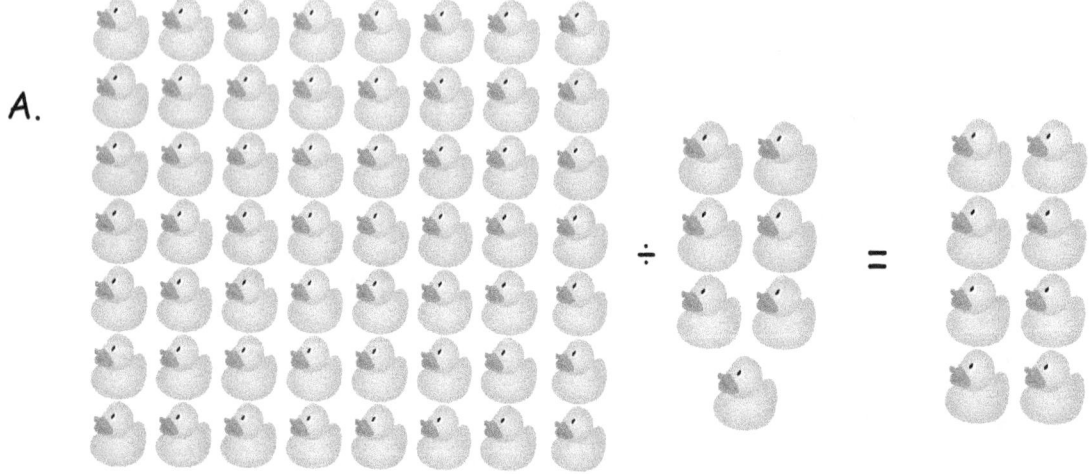

B.

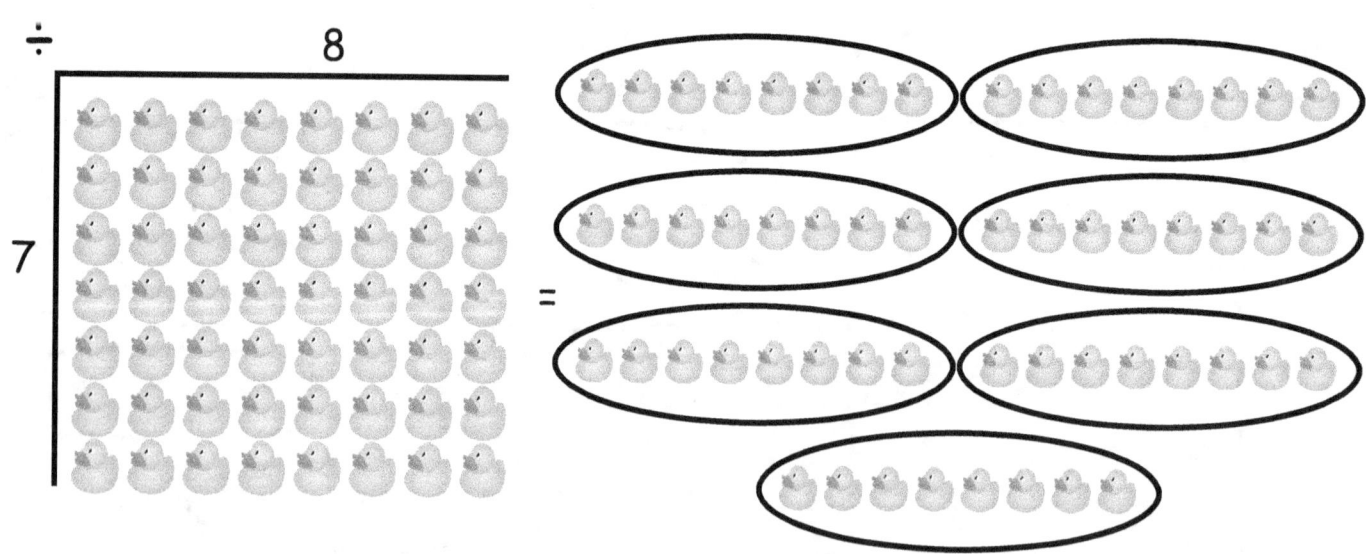

C.
$$56 \div 7 = 8$$

9. Lets learn $63 \div 7 = 9$

A. 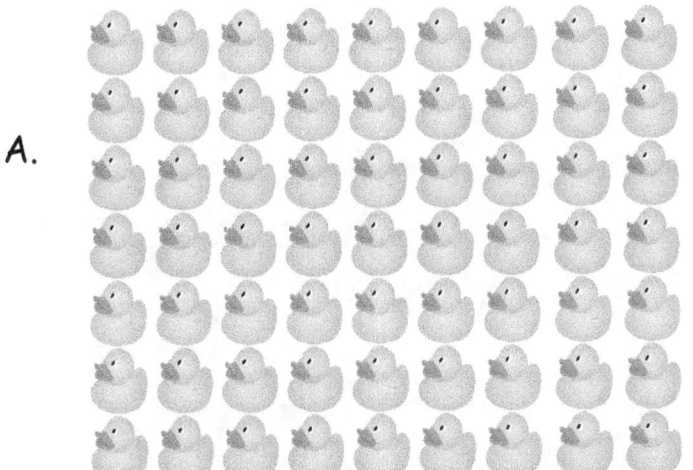 ÷

B.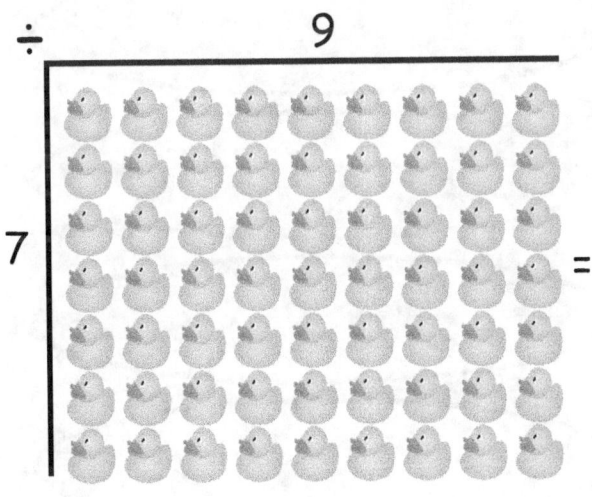

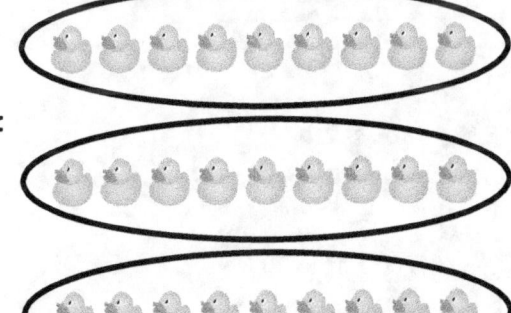

C. $63 \div 7 = 9$

www.math-knots.com

10. Lets learn $70 \div 7 = 10$

A.

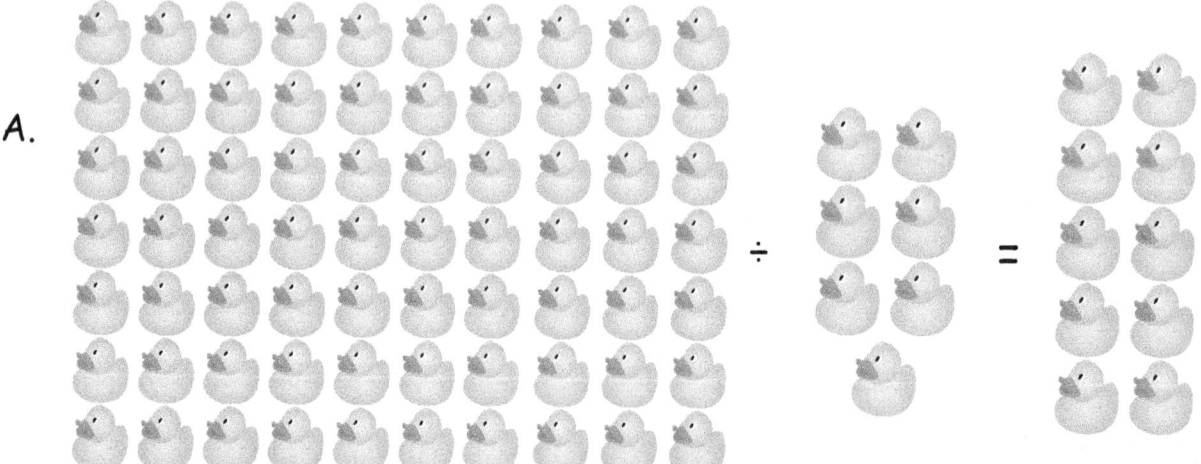

B.

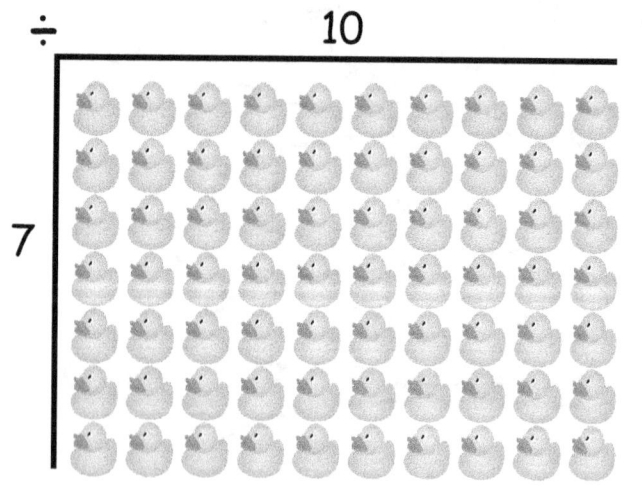

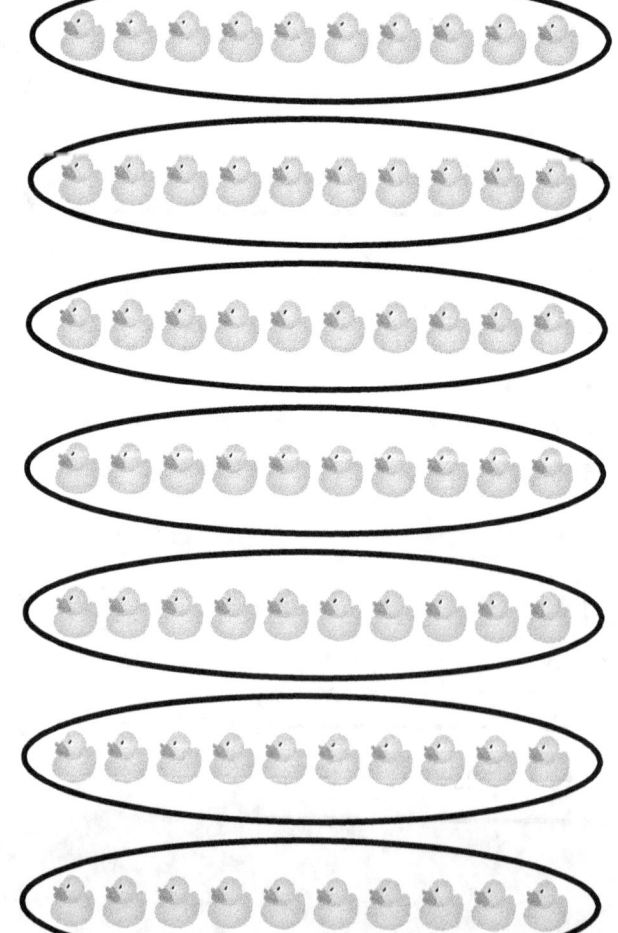

C. $\boxed{70 \div 7 = 10}$

www.math-knots.com

11. Lets learn 77 ÷ 7 = 11

A.

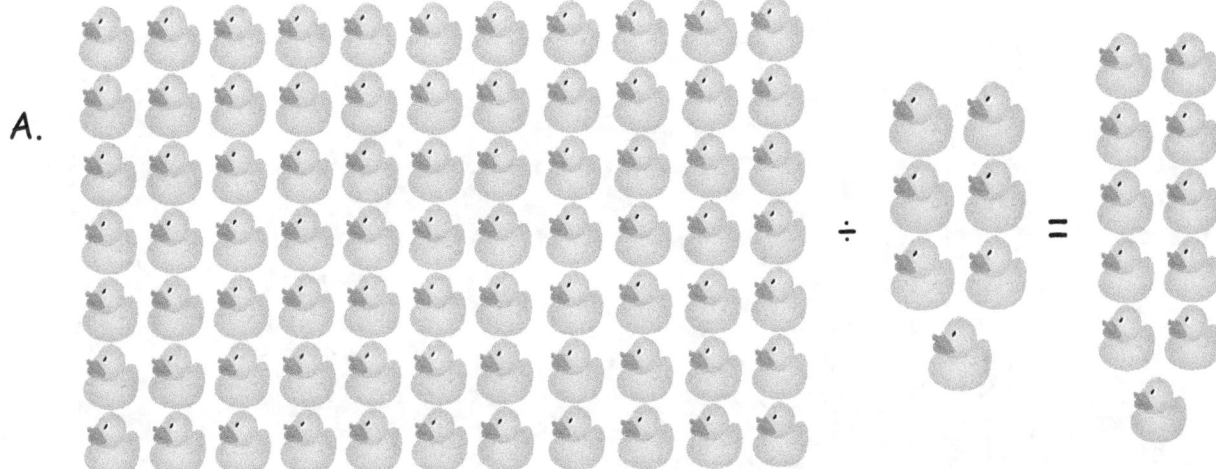

B.

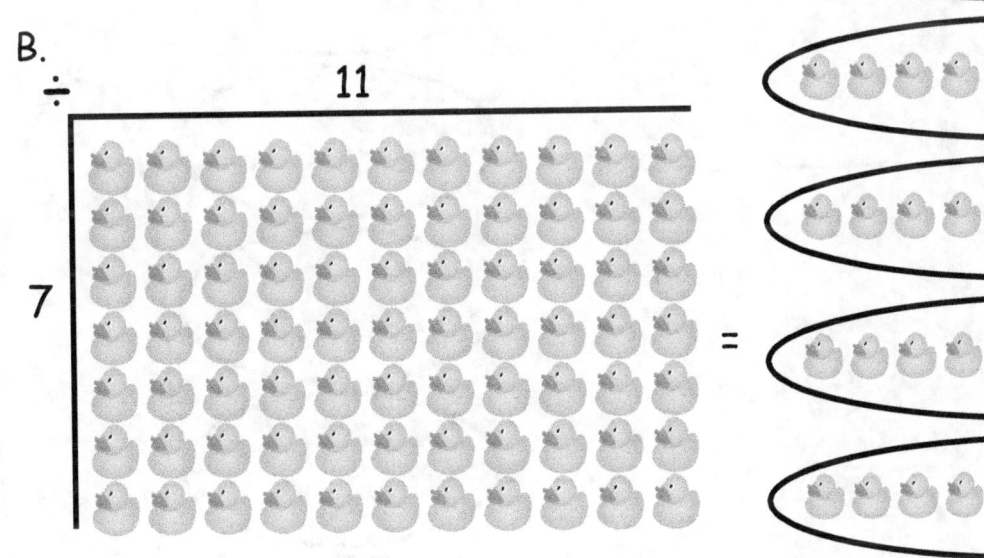

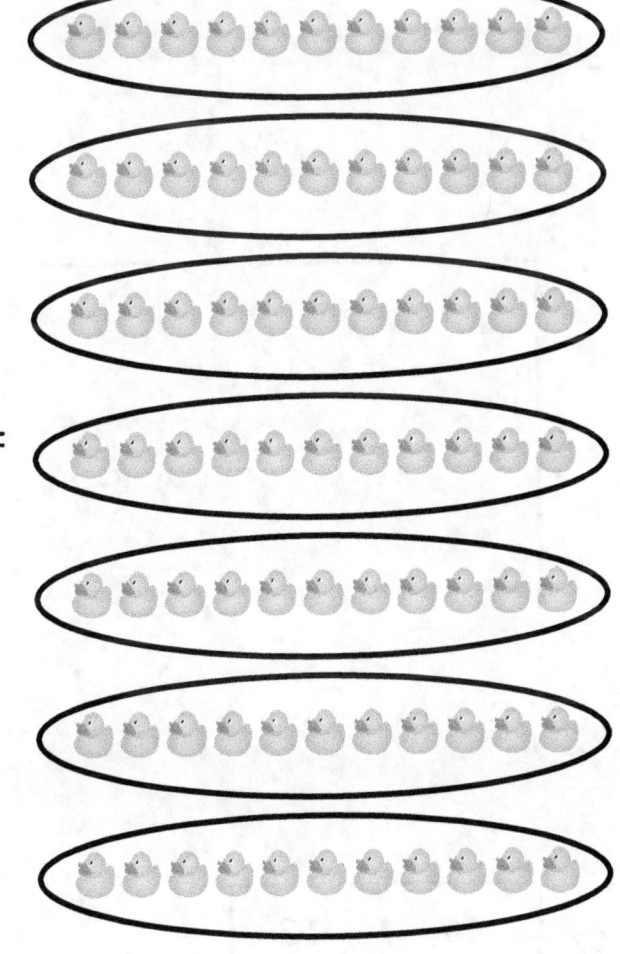

C. **77 ÷ 7 = 11**

www.math-knots.com

12. Lets learn 84 ÷ 7 = 12

A.

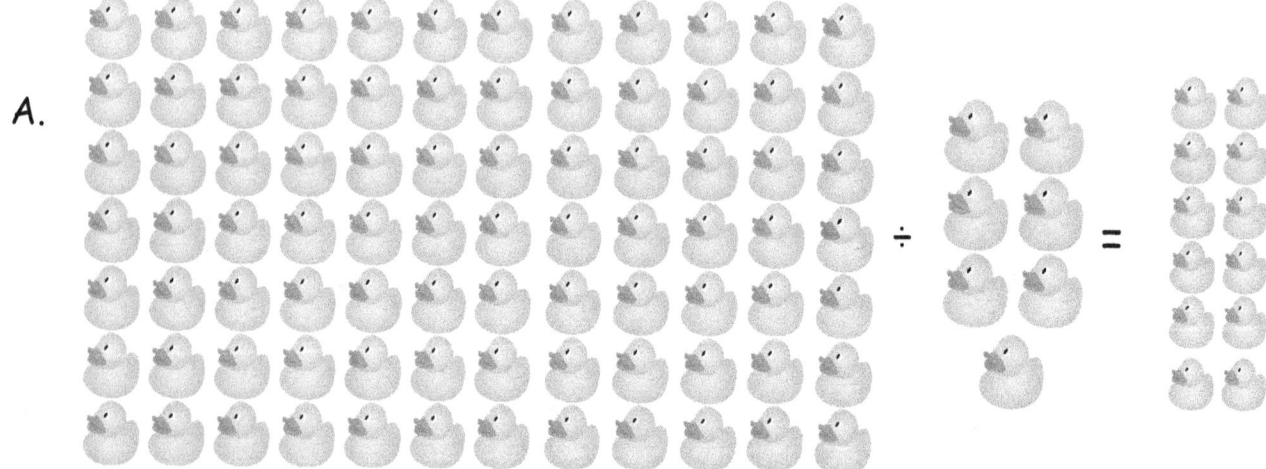

B.

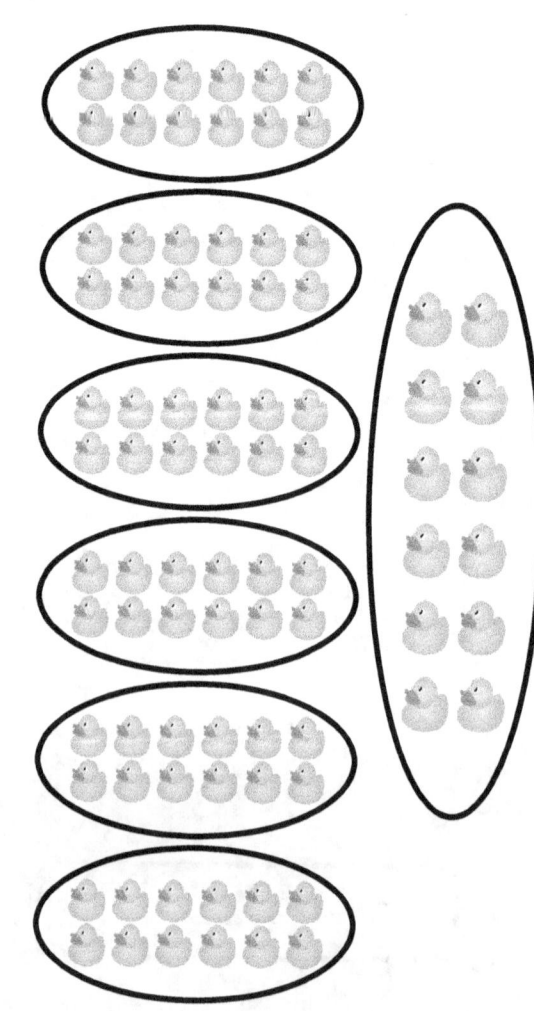

C. $84 \div 7 = 12$

www.math-knots.com

Exercise - 1

(A) $7\overline{)7}$ (F) $7\overline{)42}$ (K) $7\overline{)77}$

(B) $7\overline{)14}$ (G) $7\overline{)49}$ (L) $7\overline{)84}$

(C) $7\overline{)21}$ (H) $7\overline{)56}$ (M) $7\overline{)91}$

(D) $7\overline{)28}$ (I) $7\overline{)63}$ (N) $7\overline{)98}$

(E) $7\overline{)35}$ (J) $7\overline{)70}$ (O) $7\overline{)105}$

www.math-knots.com

Exercise - 2

1.	$7 \div 7 =$	_____
2.	$14 \div 7 =$	_____
3.	$21 \div 7 =$	_____
4.	$28 \div 7 =$	_____
5.	$35 \div 7 =$	_____
6.	$42 \div 7 =$	_____
7.	$49 \div 7 =$	_____
8.	$56 \div 7 =$	_____
9.	$63 \div 7 =$	_____
10.	$70 \div 7 =$	_____
11.	$77 \div 7 =$	_____
12.	$84 \div 7 =$	_____

1	$\times$ _____ $=$	7
2	$\times$ _____ $=$	14
3	$\times$ _____ $=$	21
4	$\times$ _____ $=$	28
5	$\times$ _____ $=$	35
6	$\times$ _____ $=$	42
7	$\times$ _____ $=$	49
8	$\times$ _____ $=$	56
9	$\times$ _____ $=$	63
10	$\times$ _____ $=$	70
11	$\times$ _____ $=$	77
12	$\times$ _____ $=$	84

Did you know division is splitting a number up by any give number.

www.math-knots.com

 # Exercise - 3

1.   I am a number, I divide myself, into one equal group of 7. What am I ?

   (A) 0   (B) 1

   (C) 14   (D) 7

2.   I am a number, I divide myself, into seven equal groups of 1. What am I ?

   (A) 1   (B) 7

   (C) 21   (D) 0

3.   I am a number, I divide myself, into seven equal groups of 2. What am I ?

   (A) 1   (B) 7

   (C) 2   (D) 14

4.   I am a number, I divide myself, into seven equal groups of 3. What am I ?

   (A) 14   (B) 4

   (C) 3   (D) 21

5.   I am a number, I divide myself, into seven equal groups of 4. What am I ?

   (A) 28   (B) 7

   (C) 4   (D) 14

6. I am a number, I divide myself, into seven equal groups of 5. What am I ?

(A) 7

(B) 1

(C) 35

(D) 21

7. I am a number, I divide myself, into seven equal groups of 6. What am I ?

(A) 6

(B) 42

(C) 7

(D) 35

8. I am a number, I divide myself, into seven equal groups of 7. What am I ?

(A) 49

(B) 7

(C) 14

(D) 42

9. I am a number, I divide myself, into seven equal groups of 8. What am I ?

(A) 8

(B) 28

(C) 7

(D) 56

10. I am a number, I divide myself, into seven equal groups of 9. What am I ?

(A) 7

(B) 9

(C) 36

(D) 63

11.   I am a number, I divide myself, into seven equal groups of 10. What am I ?

(A) 70   (B) 60

(C) 10   (D) 7

12.   I am a number, I divide myself, into seven equal groups of 11. What am I ?

(A) 66   (B) 11

(C) 77   (D) 7

13.   I am a number, I divide myself, into seven equal groups of 12. What am I ?

(A) 7   (B) 84

(C) 12   (D) 60

14.   I am a number, I divide myself, into seven equal groups of 13. What am I ?

(A) 91   (B) 42

(C) 13   (D) 7

15.   I am a number, I divide myself, into seven equal groups of 14. What am I ?

(A) 98   (B) 7

(C) 14   (D) 70

123

Exercise - 4

Solve the maze run below.

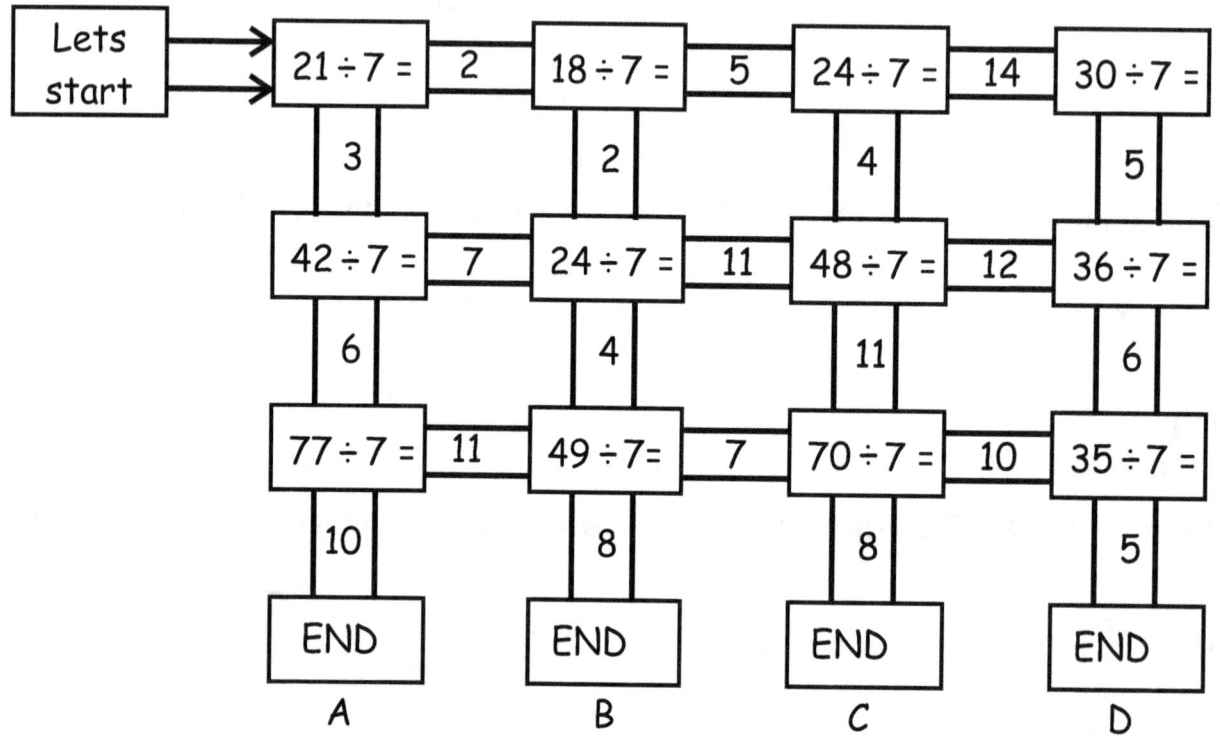

Lets start	21 ÷ 7 =	2	18 ÷ 7 =	5	24 ÷ 7 =	14	30 ÷ 7 =
	3		2		4		5
	42 ÷ 7 =	7	24 ÷ 7 =	11	48 ÷ 7 =	12	36 ÷ 7 =
	6		4		11		6
	77 ÷ 7 =	11	49 ÷ 7 =	7	70 ÷ 7 =	10	35 ÷ 7 =
	10		8		8		5
	END A		END B		END C		END D

Who won the race ? _____

www.math-knots.com

Exercise - 5

1. $7 \div \boxed{} = 1$ then $\boxed{} =$ _____

2. $14 \div \boxed{} = 7$ then $\boxed{} =$ _____

3. $21 \div \boxed{} = 7$ then $\boxed{} =$ _____

4. $28 \div \boxed{} = 7$ then $\boxed{} =$ _____

5. $35 \div \boxed{} = 7$ then $\boxed{} =$ _____

6. $42 \div \boxed{} = 7$ then $\boxed{} =$ _____

7. $49 \div \boxed{} = 7$ then $\boxed{} =$ _____

8. $56 \div \boxed{} = 7$ then $\boxed{} =$ _____

9. $63 \div \boxed{} = 7$ then $\boxed{} =$ _____

10. $70 \div \boxed{} = 7$ then $\boxed{} =$ _____

11. $77 \div \boxed{} = 7$ then $\boxed{} =$ _____

12. $84 \div \boxed{} = 7$ then $\boxed{} =$ _____

Hey you are an expert of division facts of #7 !!!

Division is opposite of Multiplication.
Division is splitting into equal parts or groups or equal sharing or equal partitioning.
Dividend: The dividend is the number that is being divided in the division process.
Divisor: The number by which dividend is being divided by is called divisor.
Quotient: A quotient is a result obtained in division process.

$$16 \div 8 = 2$$

Dividend. Divisor. Quotient
Let's learn division facts for #8

www.math-knots.com

www.math-knots.com

1. Lets learn $8 \div 1 = 8$

A.

 ÷ 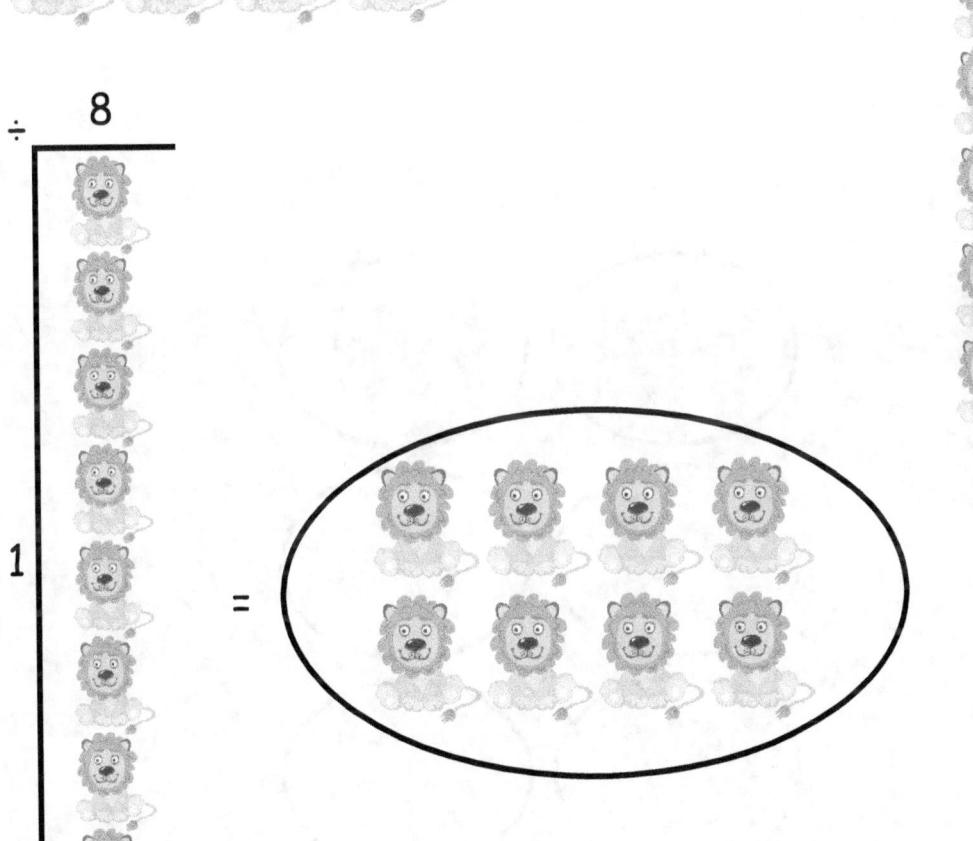 =

B.

C. $8 \div 1 = 8$

2. Lets learn 16 ÷ 8 = 2

A.

B.

÷ 2

8 =

C. 16 ÷ 8 = 2

www.math-knots.com

3. Lets learn 24 ÷ 8 = 3

A.

B.

C.

$$24 \div 8 = 3$$

4. Lets learn 32 ÷ 8 = 4

A.

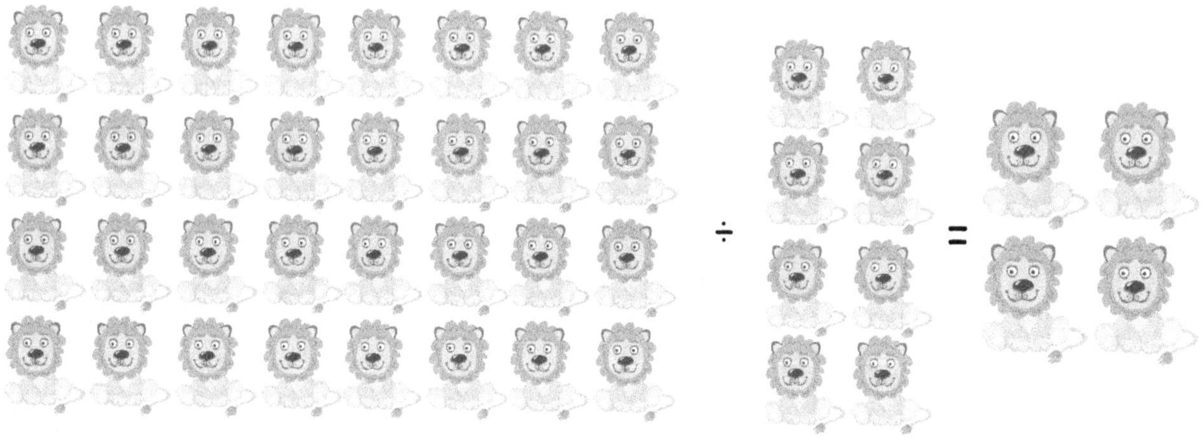

B.

÷

4

8

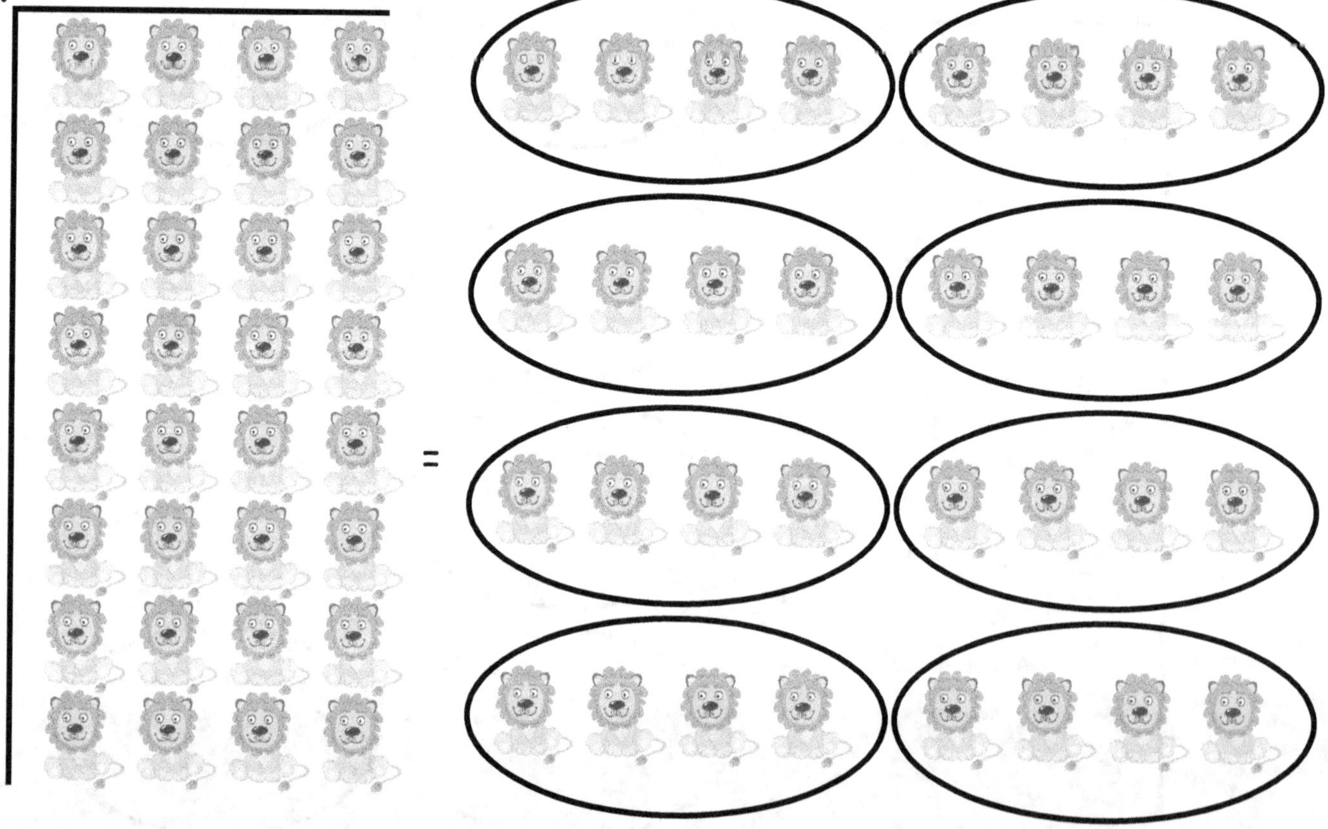

=

C. 32 ÷ 8 = 4

5. Lets learn $40 \div 8 = 5$

A.

B.

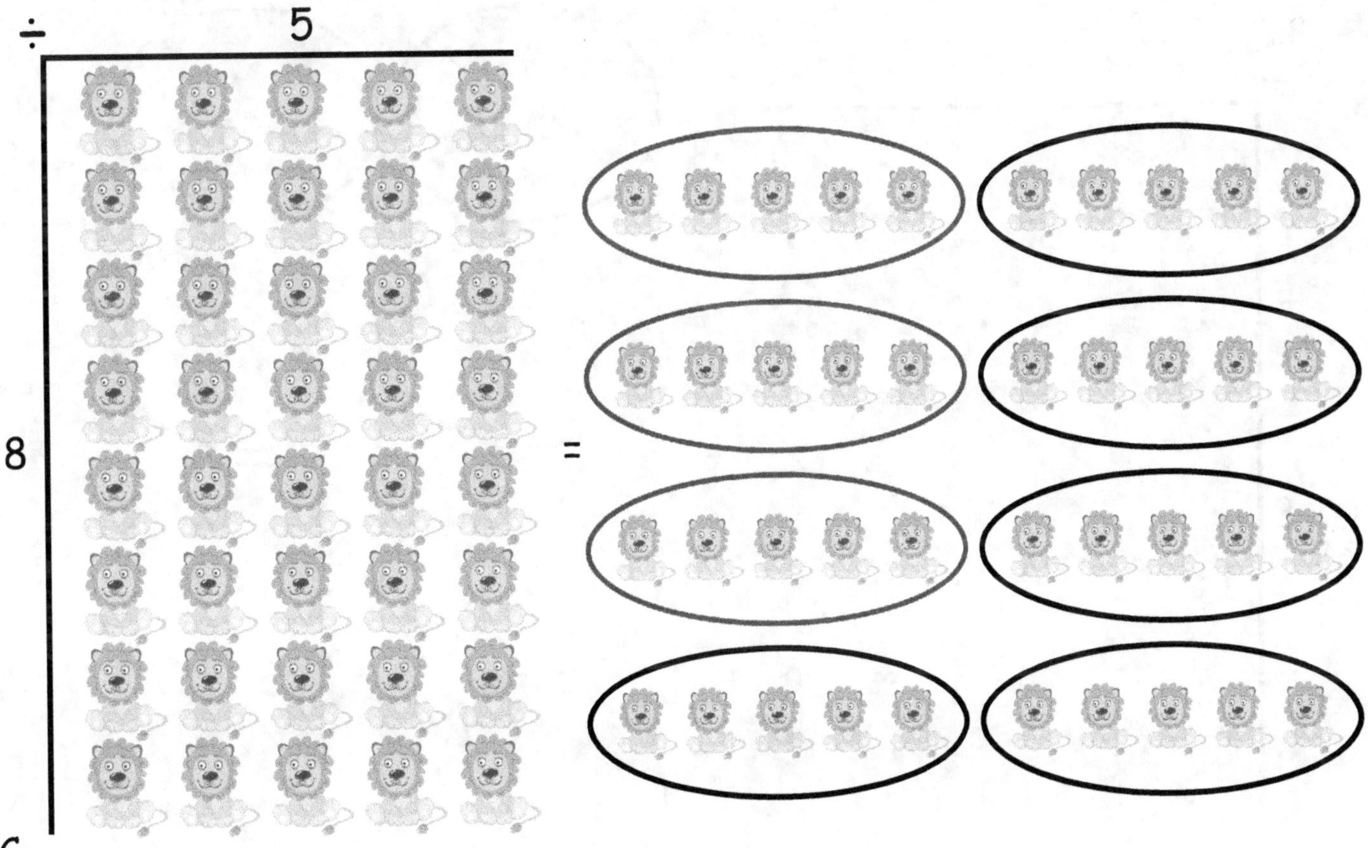

C.

$$40 \div 8 = 5$$

6.   Lets learn 48 ÷ 8 = 6

A.

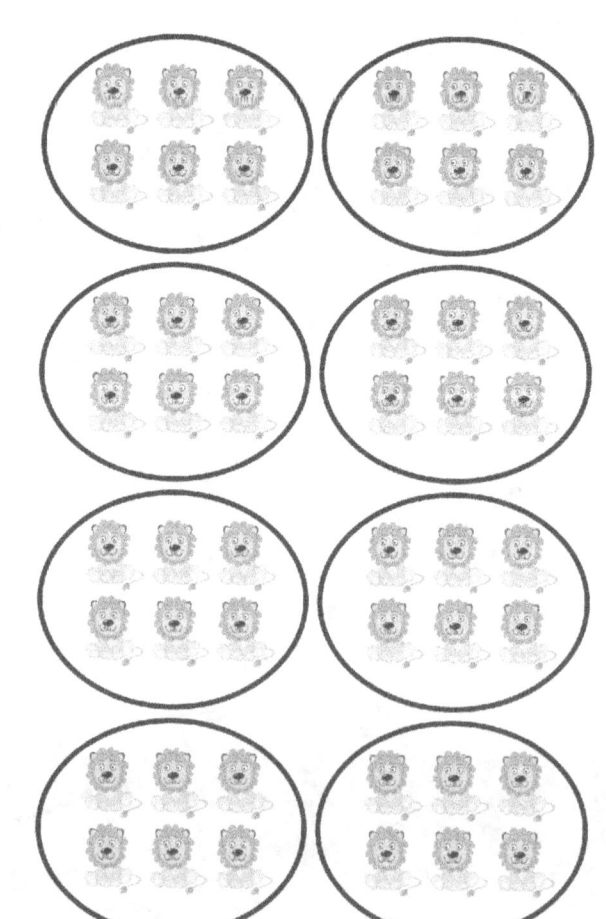

B.

C.   **48 ÷ 8 = 6**

7. Lets learn 56 ÷ 8 = 7

A.

÷

=

B. ÷

7

8

=

C.

$$56 \div 8 = 7$$

www.math-knots.com

8.   Lets learn $64 \div 8 = 8$

A.

B.

C.

$$64 \div 8 = 8$$

9.   Lets learn 72 ÷ 8 = 9

A.

B.

$$\div \quad 9$$

8

=

C. $\boxed{72 \div 8 = 9}$

10. Lets learn 80 ÷ 8 = 10

A.

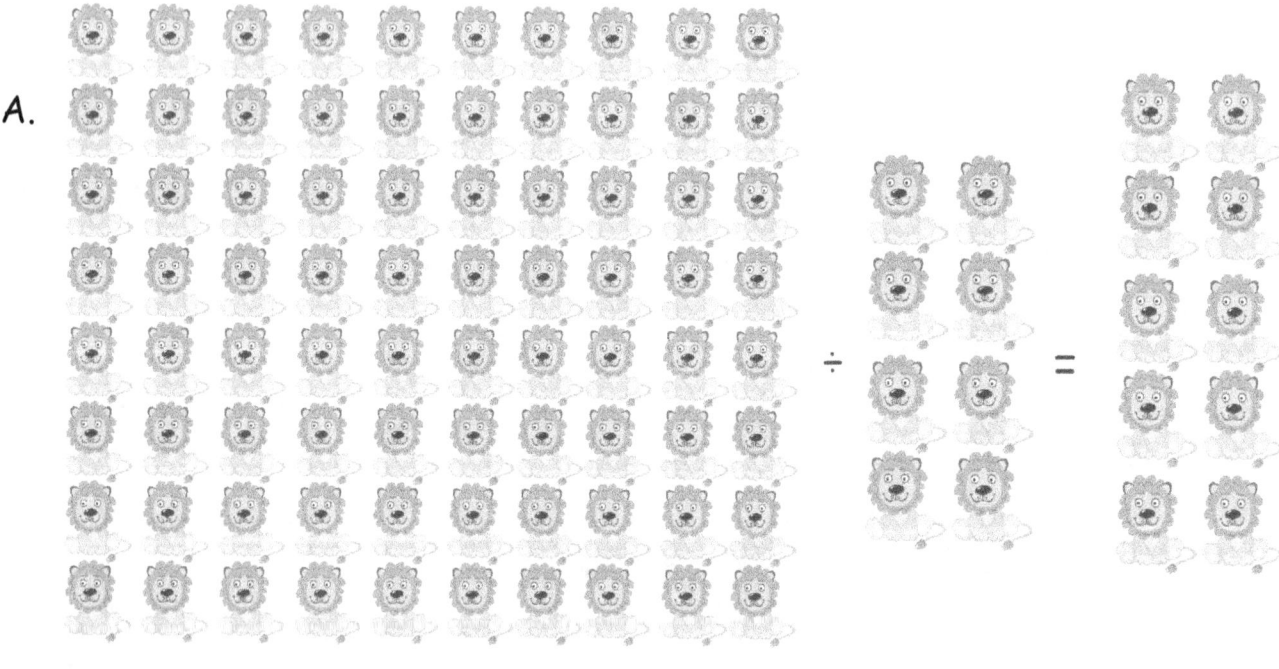

B.

$$÷$$

10

8

=

C. **80 ÷ 8 = 10**

www.math-knots.com

11.   Lets learn 88 ÷ 8 = 11

A.

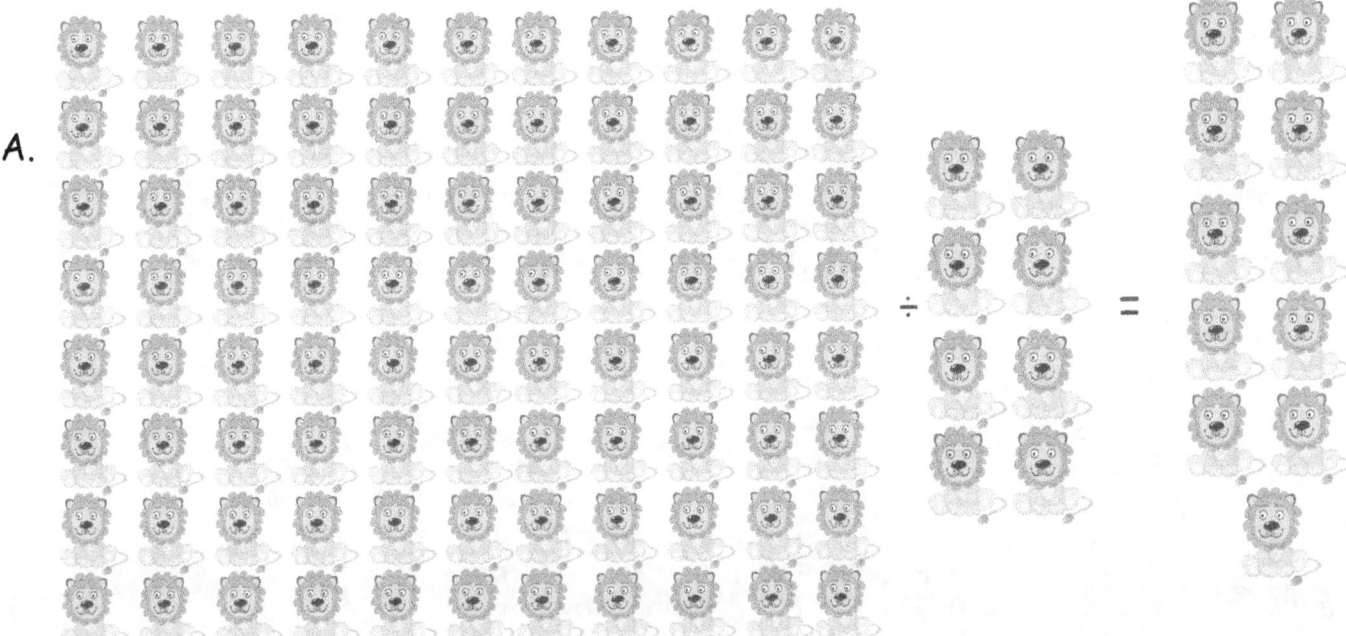

B.

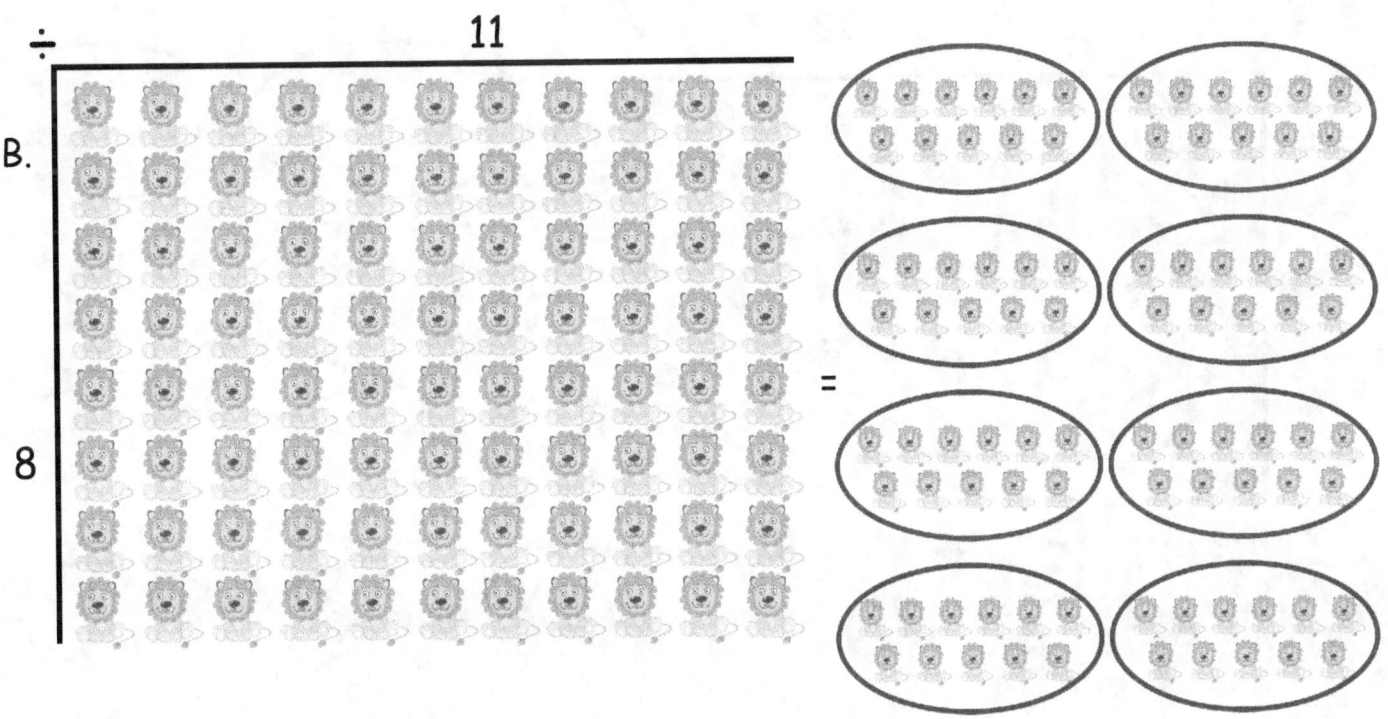

C. | 88 ÷ 8 = 11 |

www.math-knots.com

12. Lets learn 96 ÷ 8 = 12

A.

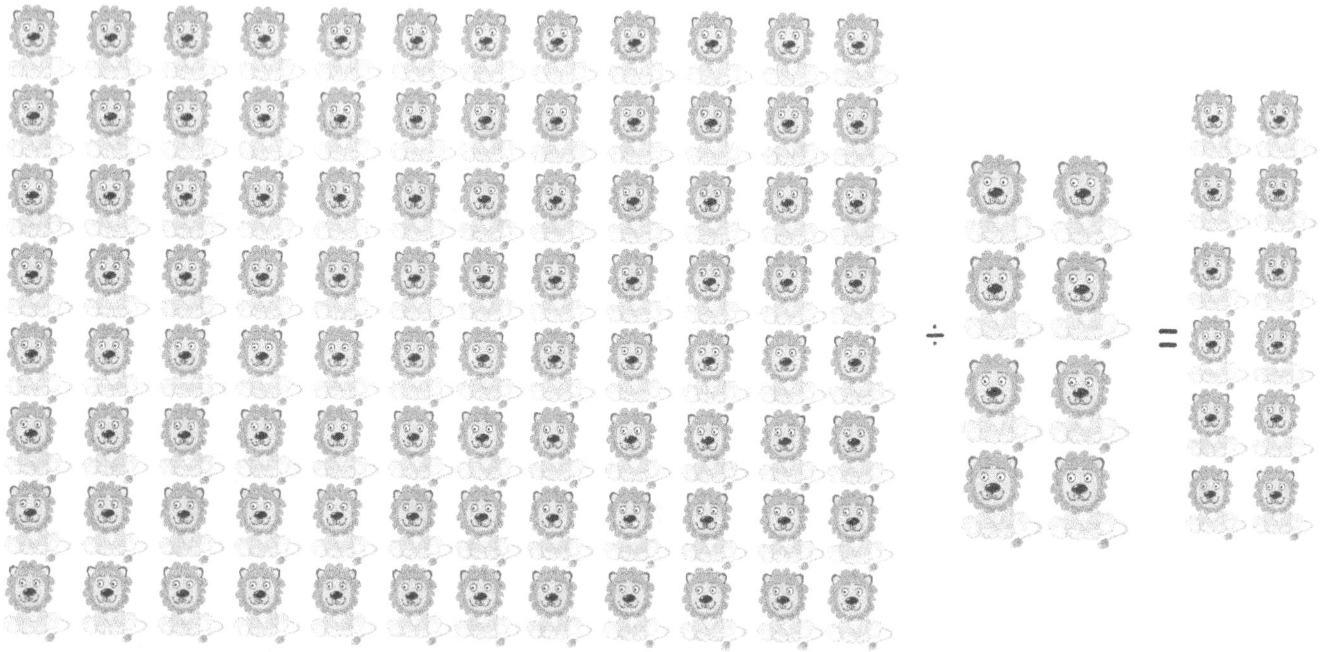

B.

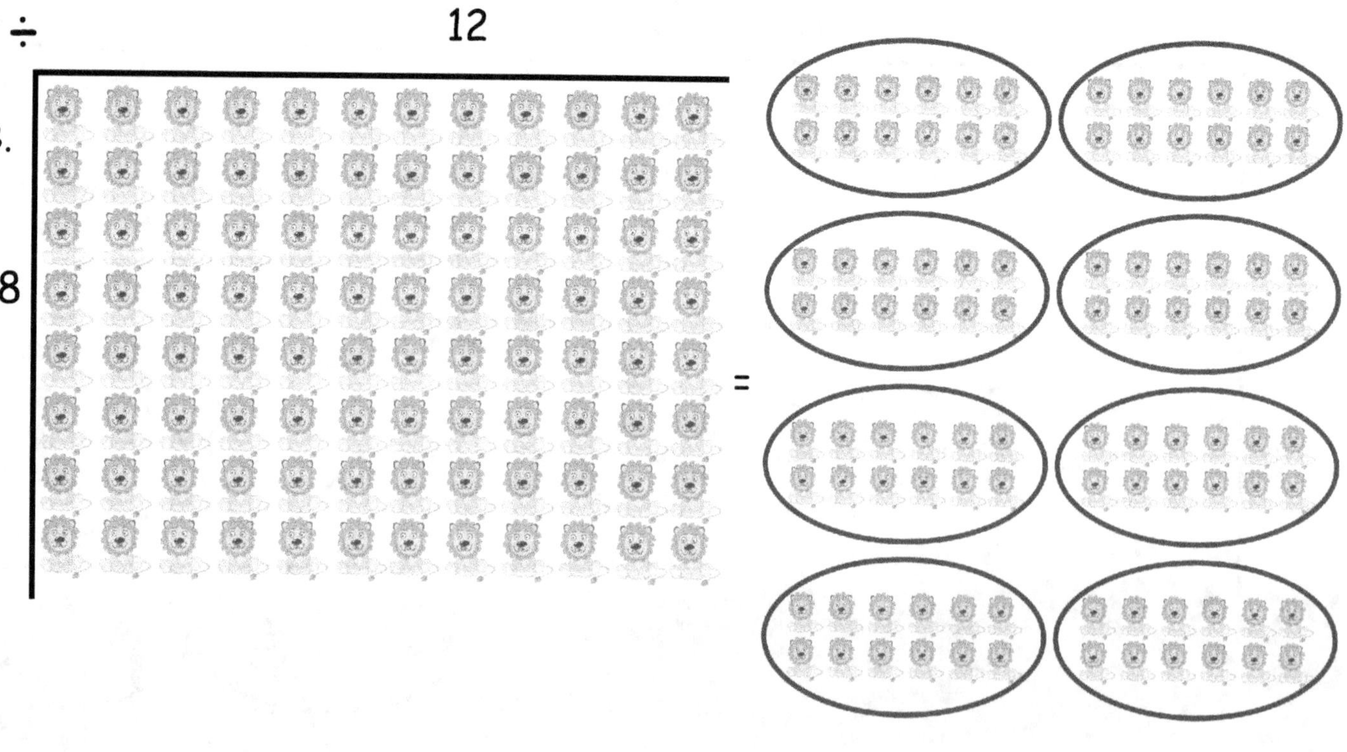

C. $\boxed{96 \div 8 = 12}$

Exercise - 1

(A) 8⟌8   (F) 8⟌48   (K) 8⟌88

(B) 8⟌16   (G) 8⟌56   (L) 8⟌96

(C) 8⟌24   (H) 8⟌64   (M) 8⟌104

(D) 8⟌32   (I) 8⟌72   (N) 8⟌112

(E) 8⟌40   (J) 8⟌80   (O) 8⟌120

www.math-knots.com

Exercise - 2

1.	8 ÷ 8 =	_____
2.	16 ÷ 8 =	_____
3.	24 ÷ 8 =	_____
4.	32 ÷ 8 =	_____
5.	40 ÷ 8 =	_____
6.	48 ÷ 8 =	_____
7.	56 ÷ 8 =	_____
8.	64 ÷ 8 =	_____
9.	72 ÷ 8 =	_____
10.	80 ÷ 8 =	_____
11.	88 ÷ 8 =	_____
12.	96 ÷ 8 =	_____

1	× _____	= 8
2	× _____	= 16
3	× _____	= 24
4	× _____	= 32
5	× _____	= 40
6	× _____	= 48
7	× _____	= 56
8	× _____	= 64
9	× _____	= 72
10	× _____	= 80
11	× _____	= 88
12	× _____	= 96

Did you know division is splitting a number up by any give number.

www.math-knots.com

 **Exercise - 3**

1. I am a number, I divide myself, into one equal group of 8. What am I ?

   (A) 0   (B) 1

   (C) 10   (D) 8

2. I am a number, I divide myself, into eight equal groups of 1. What am I ?

   (A) 1   (B) 6

   (C) 8   (D) 16

3. I am a number, I divide myself, into eight equal groups of 2. What am I ?

   (A) 16   (B) 0

   (C) 8   (D) 2

4. I am a number, I divide myself, into eight equal groups of 3. What am I ?

   (A) 8   (B) 16

   (C) 18   (D) 24

5. I am a number, I divide myself, into eight equal groups of 4. What am I ?

   (A) 16   (B) 32

   (C) 8   (D) 4

6. I am a number, I divide myself, into eight equal groups of 5. What am I ?

    (A) 5   (B) 8

    (C) 32   (D) 40

7. I am a number, I divide myself, into eight equal groups of 6. What am I ?

    (A) 48   (B) 6

    (C) 24   (D) 8

8. I am a number, I divide myself, into eight equal groups of 7. What am I ?

    (A) 8   (B) 24

    (C) 56   (D) 7

9. I am a number, I divide myself, into eight equal groups of 8. What am I ?

    (A) 18   (B) 8

    (C) 64   (D) 48

10. I am a number, I divide myself, into eight equal groups of 9. What am I ?

    (A) 36   (B) 9

    (C) 28   (D) 72

11. I am a number, I divide myself, into eight equal groups of 10. What am I ?

  (A) 80   (B) 60

  (C) 24   (D) 36

12. I am a number, I divide myself, into eight equal groups of 11. What am I ?

  (A) 66   (B) 48

  (C) 11   (D) 88

13. I am a number, I divide myself, into eight equal groups of 12. What am I ?

  (A) 12   (B) 96

  (C) 72   (D) 60

14. I am a number, I divide myself, into eight equal groups of 13. What am I ?

  (A) 13   (B) 104

  (C) 65   (D) 78

15. I am a number, I divide myself, into eight equal groups of 14. What am I ?

  (A) 36   (B) 14

  (C) 112   (D) 70

Exercise - 4

Solve the maze run below.

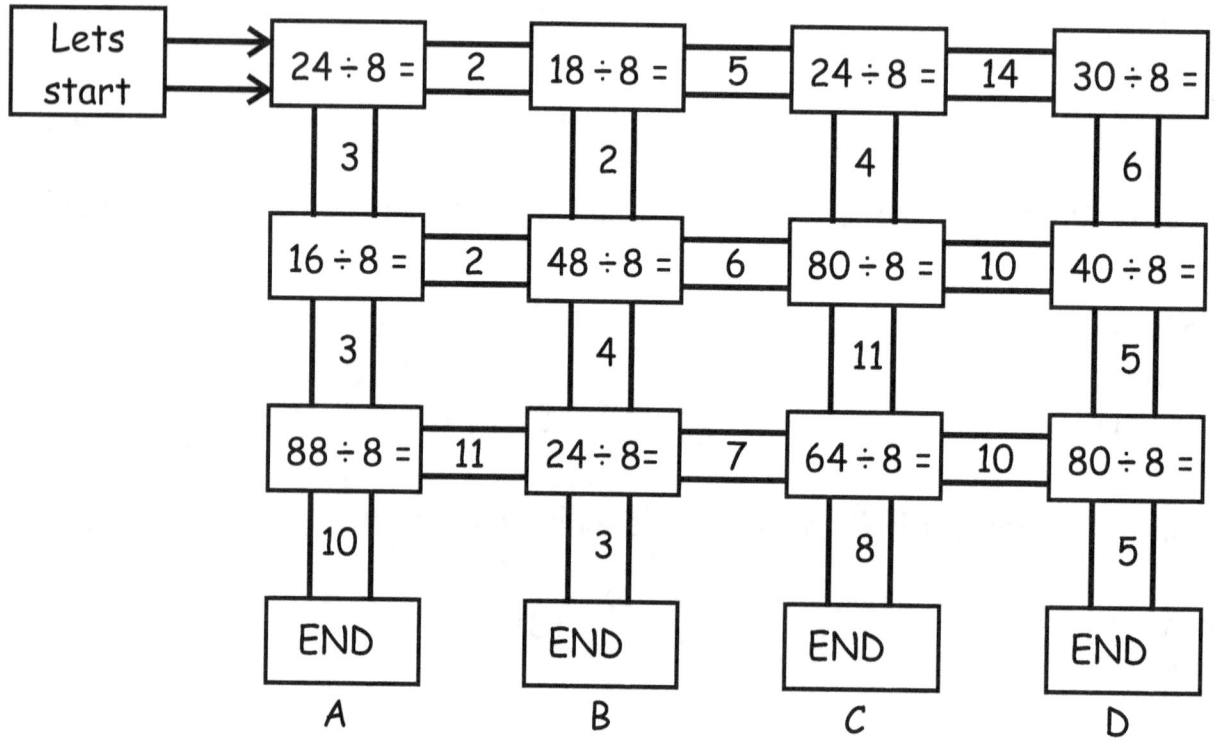

Lets start →	24 ÷ 8 =	2	18 ÷ 8 =	5	24 ÷ 8 =	14	30 ÷ 8 =

Lets start

24 ÷ 8 = 2 18 ÷ 8 = 5 24 ÷ 8 = 14 30 ÷ 8 =

3   2   4   6

16 ÷ 8 = 2 48 ÷ 8 = 6 80 ÷ 8 = 10 40 ÷ 8 =

3   4   11   5

88 ÷ 8 = 11 24 ÷ 8= 7 64 ÷ 8 = 10 80 ÷ 8 =

10   3   8   5

END   END   END   END

A   B   C   D

Who won the race ? _____

www.math-knots.com

Exercise - 5

1.   8 ÷ ☐ = 1   then   ☐ = _____

2.   16 ÷ ☐ = 8   then   ☐ = _____

3.   24 ÷ ☐ = 8   then   ☐ = _____

4.   32 ÷ ☐ = 8   then   ☐ = _____

5.   40 ÷ ☐ = 8   then   ☐ = _____

6.   48 ÷ ☐ = 8   then   ☐ = _____

7.   56 ÷ ☐ = 8   then   ☐ = _____

8.   64 ÷ ☐ = 8   then   ☐ = _____

9.   72 ÷ ☐ = 8   then   ☐ = _____

10.   80 ÷ ☐ = 8   then   ☐ = _____

11.   88 ÷ ☐ = 8   then   ☐ = _____

12.   96 ÷ ☐ = 8   then   ☐ = _____

Hey you are an expert of division facts of #8 !!!

www.math-knots.com

Division is opposite of Multiplication.
Division is splitting into equal parts or groups or equal sharing or equal partitioning.
Dividend: The dividend is the number that is being divided in the division process.
Divisor: The number by which dividend is being divided by is called divisor.
Quotient: A quotient is a result obtained in division process.

$$18 \div 9 = 2$$

Dividend. Divisor. Quotient
Let's learn division facts for #9

www.math-knots.com

www.math-knots.com

1. Lets learn 9 ÷ 1 = 9

A.  ÷ 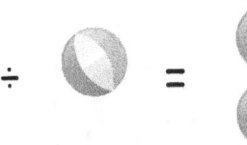 =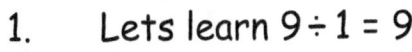

B. 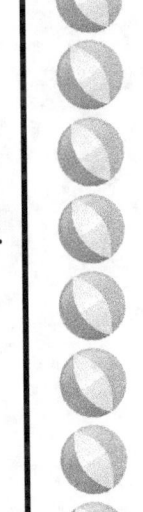 =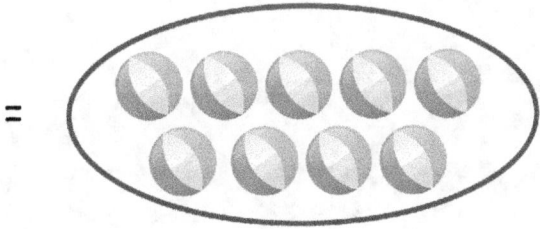

C. $9 ÷ 1 = 9$

2. Lets learn 18 ÷ 9 = 2

A.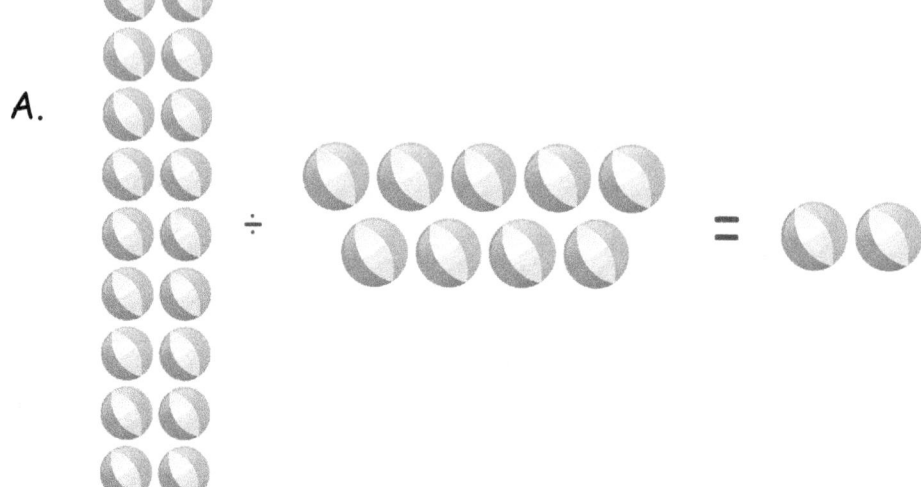

B.

$$\div \quad 2$$

9

=

C. 18 ÷ 9 = 2

3. Lets learn 27 ÷ 9 = 3

A.

 ÷ (balls) = ● ● ●

B.

÷ 3

9 =

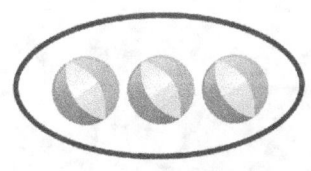 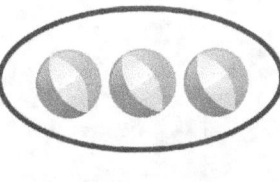 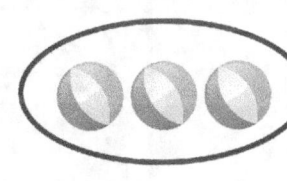

 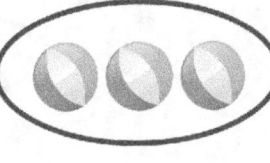 

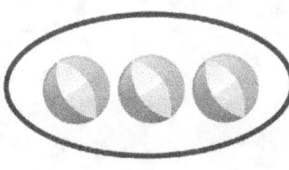 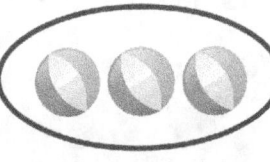 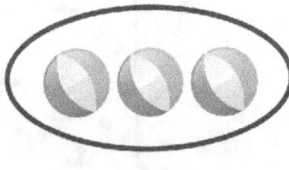

C.

| 27 ÷ 9 = 3 |

4. Lets learn 36 ÷ 9 = 4

A.

 ÷  =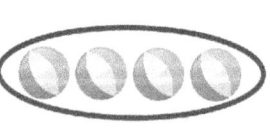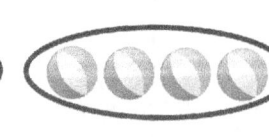

B.

÷   4

9

=

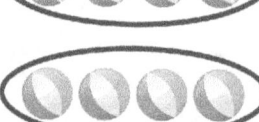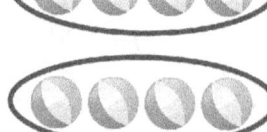

C. **36 ÷ 9 = 4**

5. Lets learn $45 \div 9 = 5$

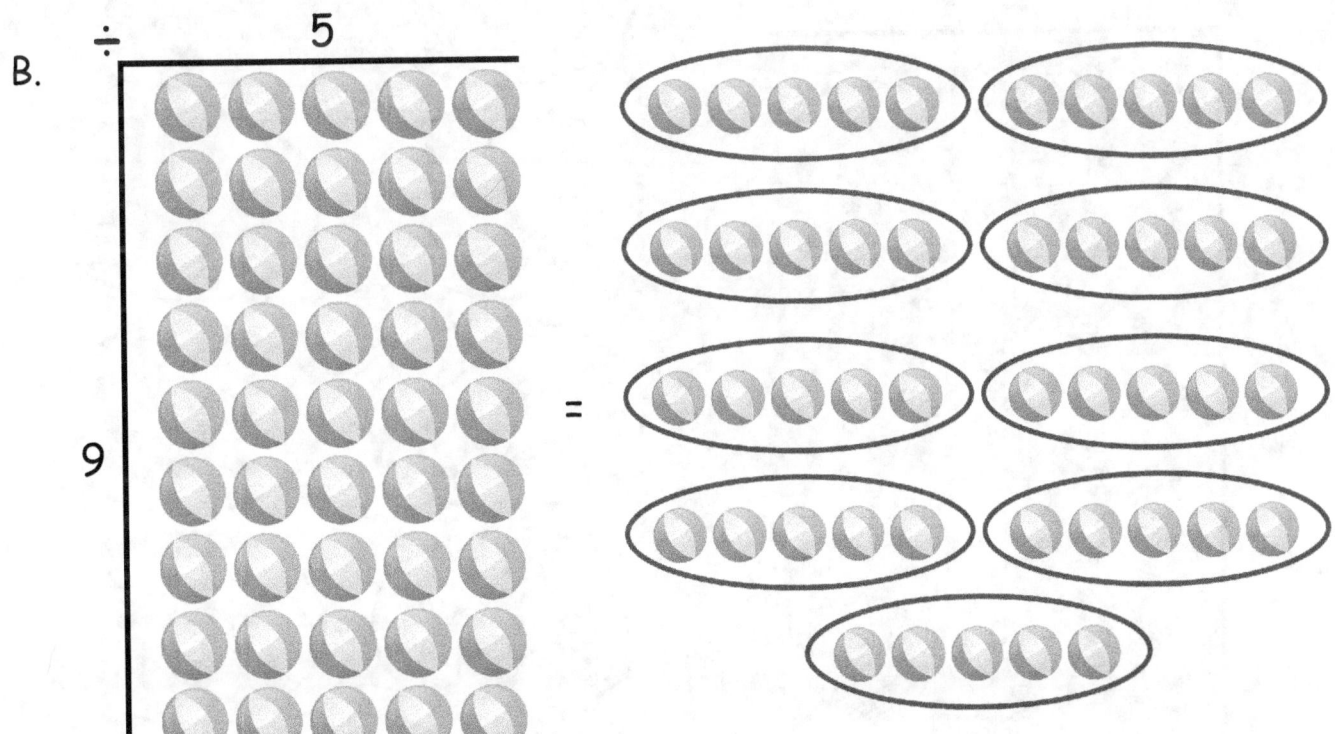

A.

B.

C. $\boxed{45 \div 9 = 5}$

6.   Lets learn 54 ÷ 9 = 6

A.

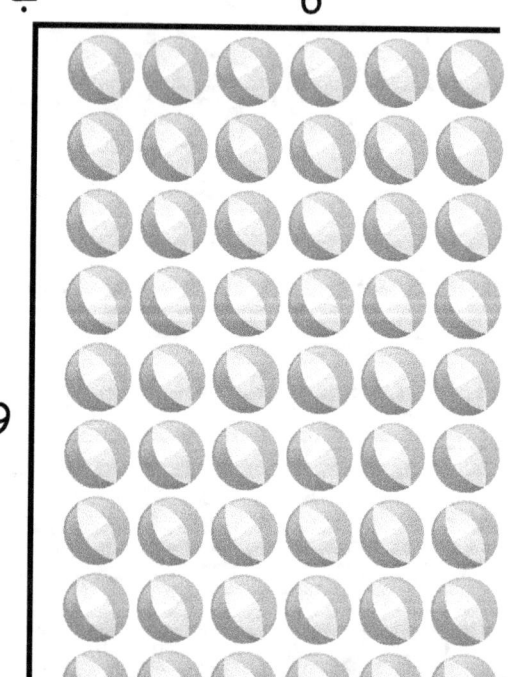 ÷ 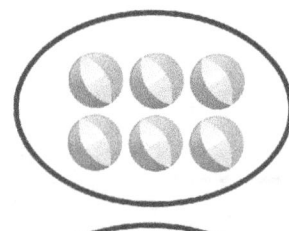 =

B.

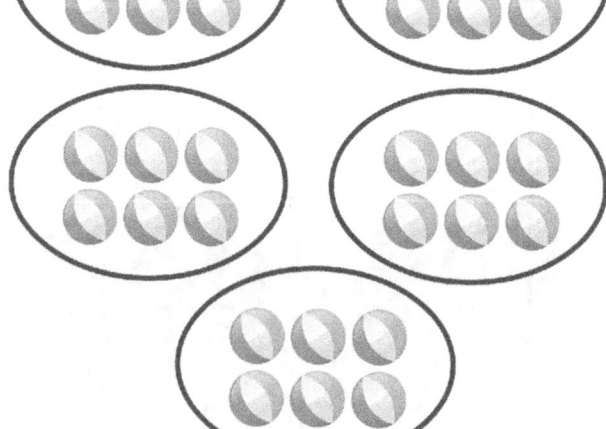

C. $54 \div 9 = 6$

www.math-knots.com

7. Lets learn $63 \div 9 = 7$

A.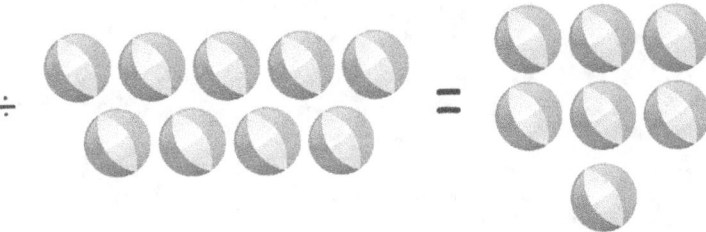

B. 7

9

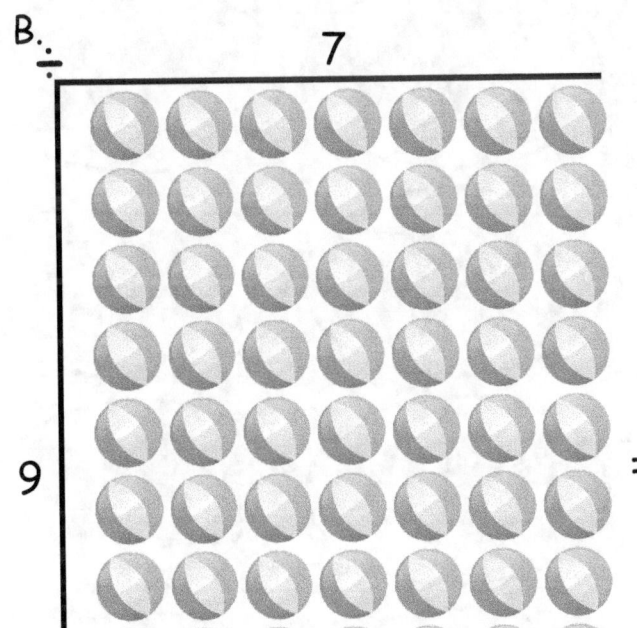 =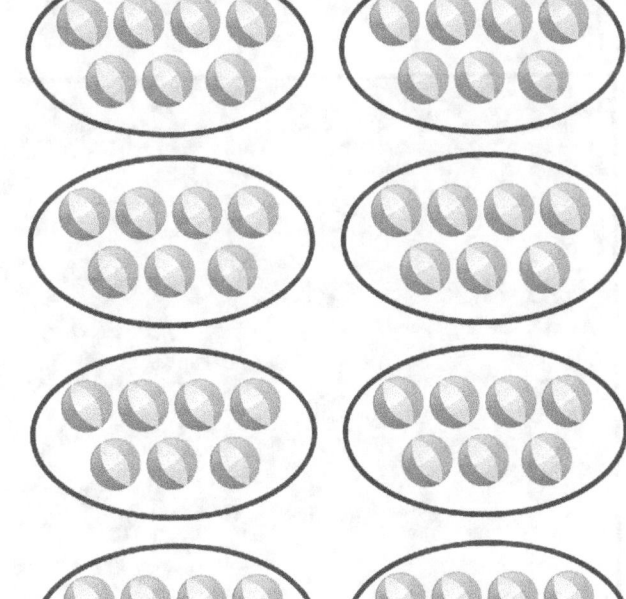

C.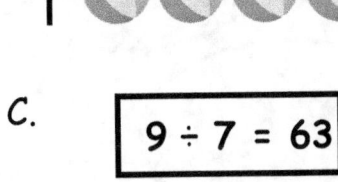

$9 \div 7 = 63$

8.   Lets learn $72 \div 9 = 8$

A.

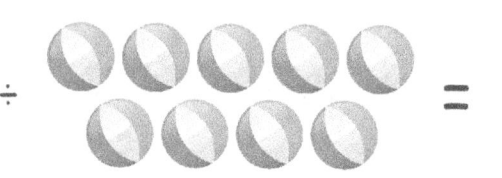

B.

$\div$

8

9

=

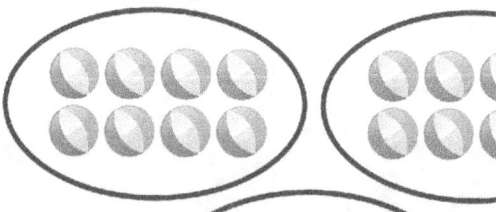

C.

$$72 \div 9 = 8$$

www.math-knots.com

9.   Lets learn $81 \div 9 = 9$

A.

B.

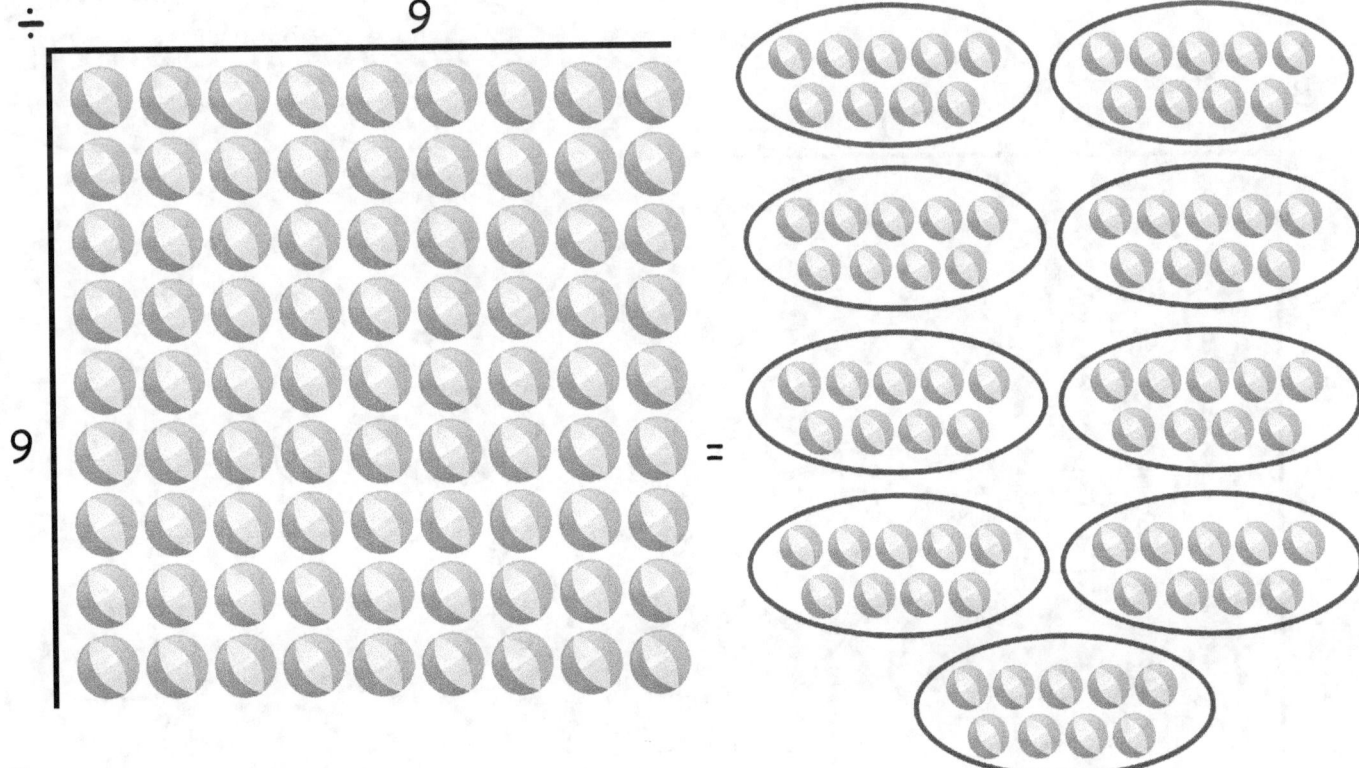

C. $81 \div 9 = 9$

10. Lets learn 90 ÷ 9 = 10

A.

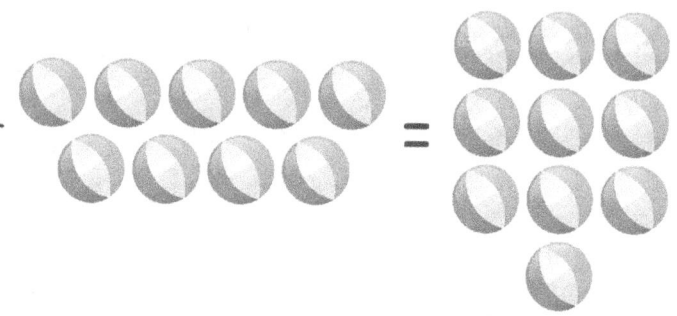

B.

÷

10

9

=

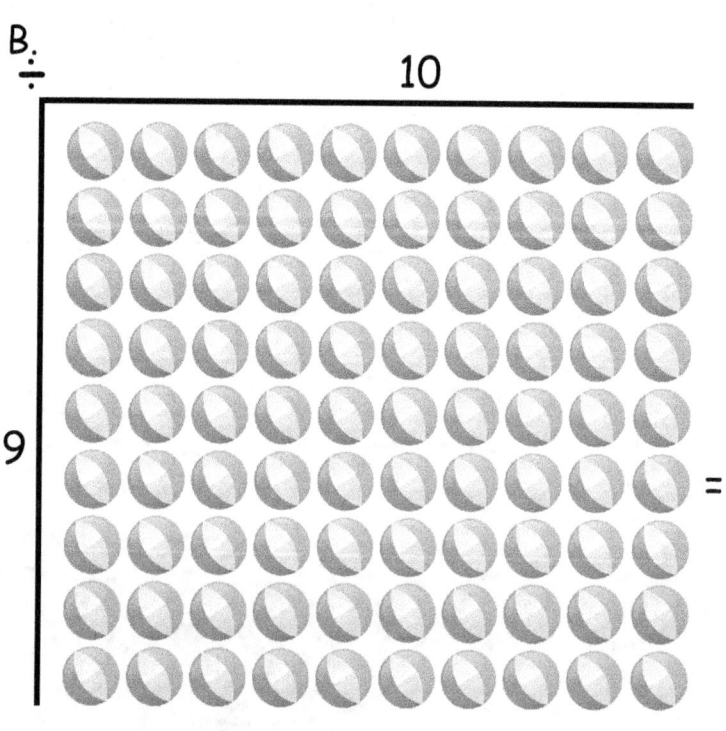

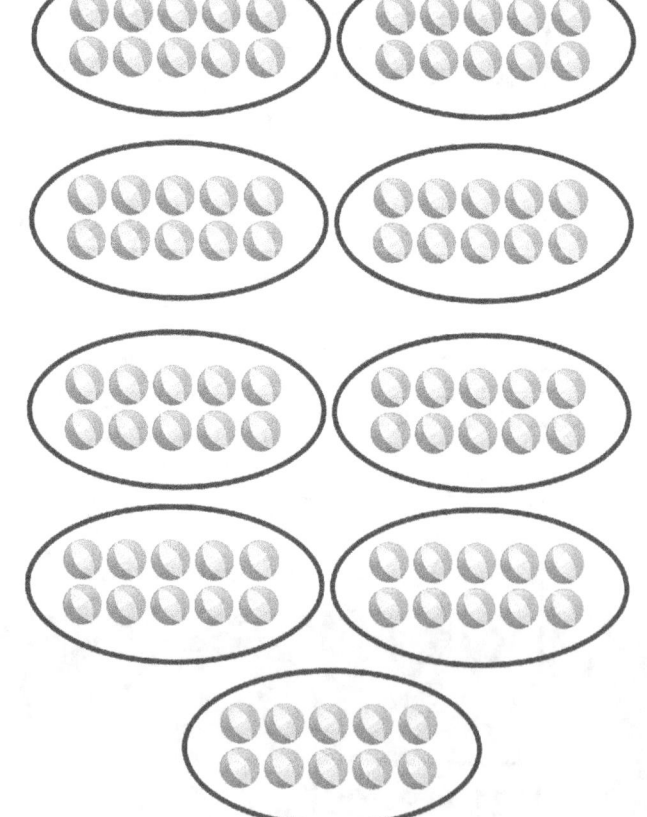

C. 90 ÷ 9 = 10

www.math-knots.com

11. Lets learn 99 ÷ 9 = 11

A.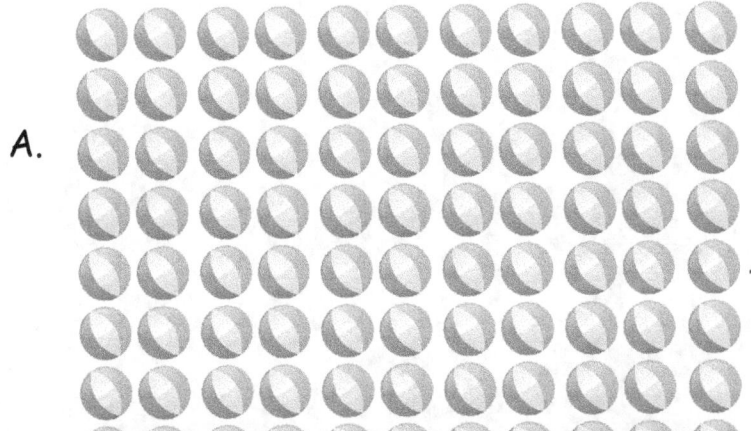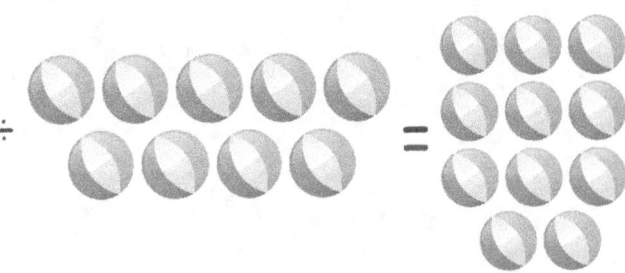

B.

÷

11

9

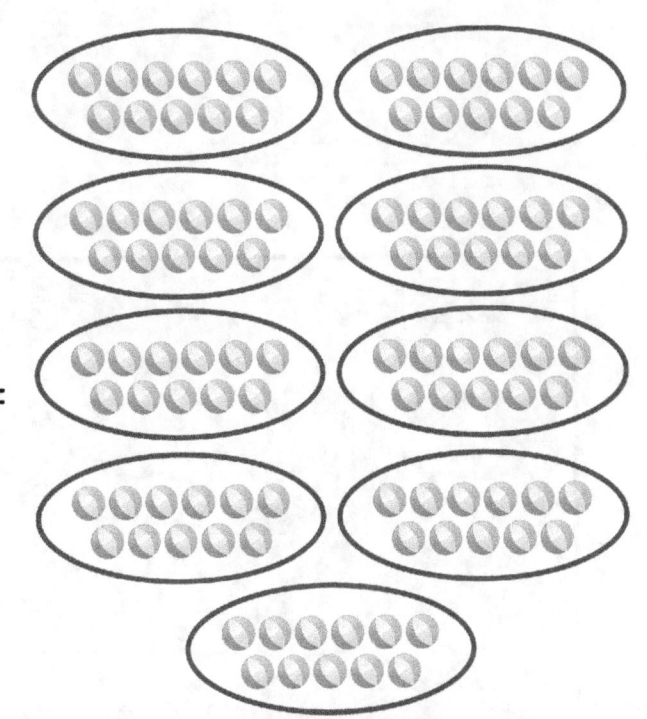

C.

$$99 \div 9 = 11$$

12. Lets learn 108 ÷ 9 = 12

A.

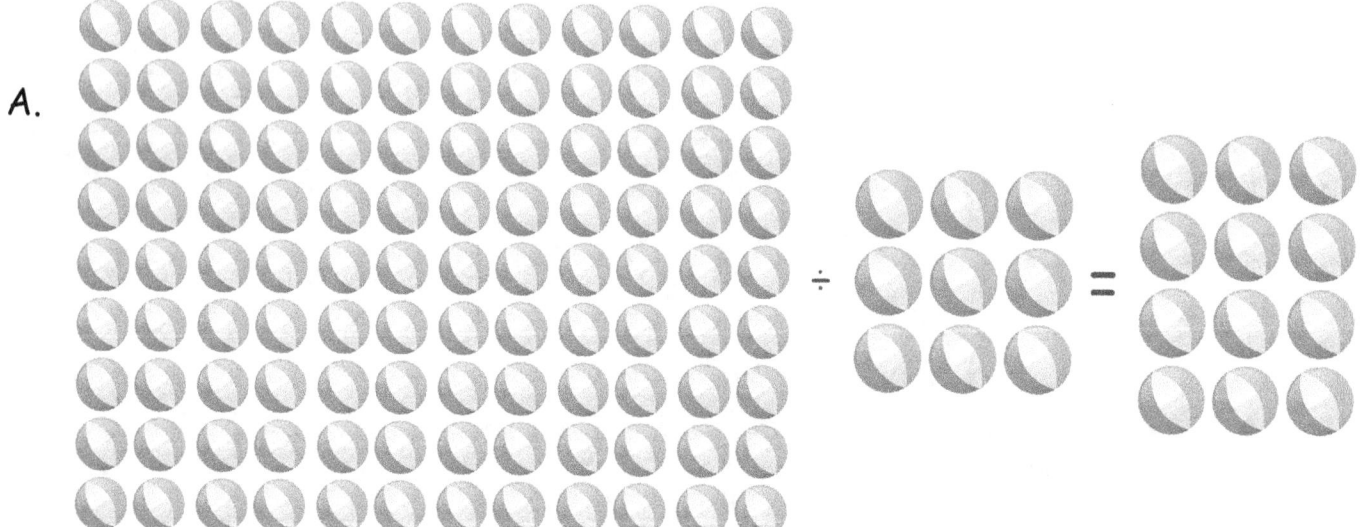

B.

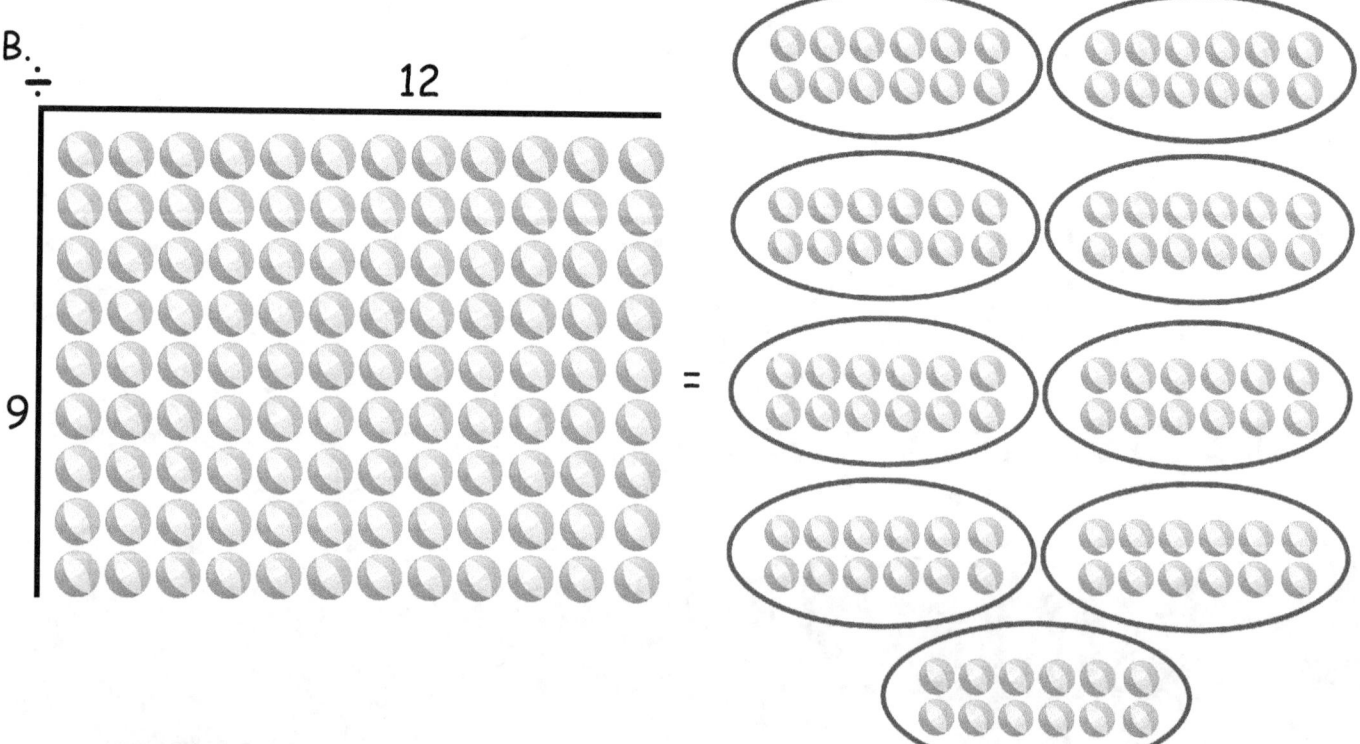

C. $108 \div 9 = 12$

Exercise - 1

(A) $9\overline{)9}$ (F) $9\overline{)54}$ (K) $9\overline{)99}$

(B) $9\overline{)18}$ (G) $9\overline{)63}$ (L) $9\overline{)108}$

(C) $9\overline{)27}$ (H) $9\overline{)72}$ (M) $9\overline{)117}$

(D) $9\overline{)36}$ (I) $9\overline{)81}$ (N) $9\overline{)126}$

(E) $9\overline{)45}$ (J) $9\overline{)90}$ (O) $9\overline{)135}$

www.math-knots.com

Exercise - 2

1.	9 ÷ 9 =	_____	1 × ____ = 9
2.	18 ÷ 9 =	_____	2 × ____ = 18
3.	27 ÷ 9 =	_____	3 × ____ = 27
4.	36 ÷ 9 =	_____	4 × ____ = 36
5.	45 ÷ 9 =	_____	5 × ____ = 45
6.	54 ÷ 9 =	_____	6 × ____ = 54
7.	63 ÷ 9 =	_____	7 × ____ = 63
8.	72 ÷ 9 =	_____	8 × ____ = 72
9.	81 ÷ 9 =	_____	9 × ____ = 81
10.	90 ÷ 9 =	_____	10 × ____ = 90
11.	99 ÷ 9 =	_____	11 × ____ = 99
12.	108 ÷ 9 =	_____	12 × ____ = 108

Did You Know...?

Did you know division is splitting a
number up by any give number.

www.math-knots.com

 **Exercise - 3**

1. I am a number, I divide myself, into one equal group of 9. What am I ?

    (A) 9   (B) 10

    (C) 1   (D) 18

2. I am a number, I divide myself, into nine equal group of 1. What am I ?

    (A) 1   (B) 27

    (C) 0   (D) 9

3. I am a number, I divide myself, into nine equal group of 2. What am I ?

    (A) 18   (B) 15

    (C) 9   (D) 2

4. I am a number, I divide myself, into nine equal group of 3. What am I ?

    (A) 21   (B) 3

    (C) 27   (D) 18

5. I am a number, I divide myself, into nine equal group of 4. What am I ?

    (A) 27   (B) 63

    (C) 36   (D) 45

www.math-knots.com

6. I am a number, I divide myself, into nine equal group of 5. What am I ?

    (A) 18   (B) 45

    (C) 36   (D) 5

7. I am a number, I divide myself, into nine equal group of 6. What am I ?

    (A) 19   (B) 27

    (C) 54   (D) 6

8. I am a number, I divide myself, into nine equal group of 7. What am I ?

    (A) 45   (B) 72

    (C) 63   (D) 7

9. I am a number, I divide myself, into nine equal group of 8. What am I ?

    (A) 72   (B) 54

    (C) 45   (D) 8

10. I am a number, I divide myself, into nine equal group of 9. What am I ?

    (A) 9   (B) 45

    (C) 81   (D) 36

www.math-knots.com

11. I am a number, I divide myself, into nine equal group of 10. What am I ?

(A) 10   (B) 50

(C) 54   (D) 90

12. I am a number, I divide myself, into nine equal group of 11. What am I ?

(A) 45   (B) 99

(C) 11   (D) 33

13. I am a number, I divide myself, into nine equal group of 12. What am I ?

(A) 108   (B) 60

(C) 91   (D) 12

14. I am a number, I divide myself, into nine equal group of 13. What am I ?

(A) 117   (B) 21

(C) 13   (D) 63

15. I am a number, I divide myself, into nine equal group of 14. What am I ?

(A) 70   (B) 14

(C) 108   (D) 126

167   www.math-knots.com

Exercise - 4

Solve the maze run below.

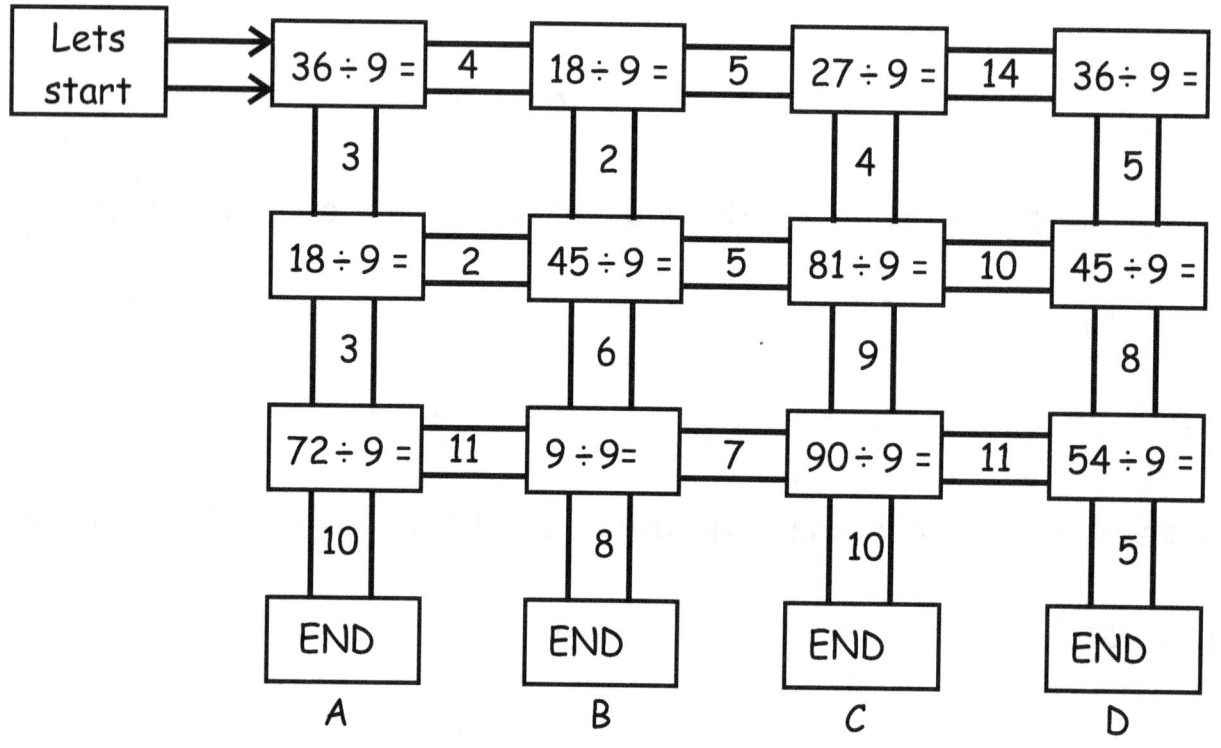

Who won the race ? _____

Exercise - 5

1. $9 \div \square = 1$ then $\square = $ _____

2. $18 \div \square = 9$ then $\square = $ _____

3. $27 \div \square = 9$ then $\square = $ _____

4. $36 \div \square = 9$ then $\square = $ _____

5. $45 \div \square = 9$ then $\square = $ _____

6. $54 \div \square = 9$ then $\square = $ _____

7. $63 \div \square = 9$ then $\square = $ _____

8. $72 \div \square = 9$ then $\square = $ _____

9. $81 \div \square = 9$ then $\square = $ _____

10. $90 \div \square = 9$ then $\square = $ _____

11. $99 \div \square = 9$ then $\square = $ _____

12. $108 \div \square = 9$ then $\square = $ _____

Hey you are an expert of division facts of #9 !!!

www.math-knots.com

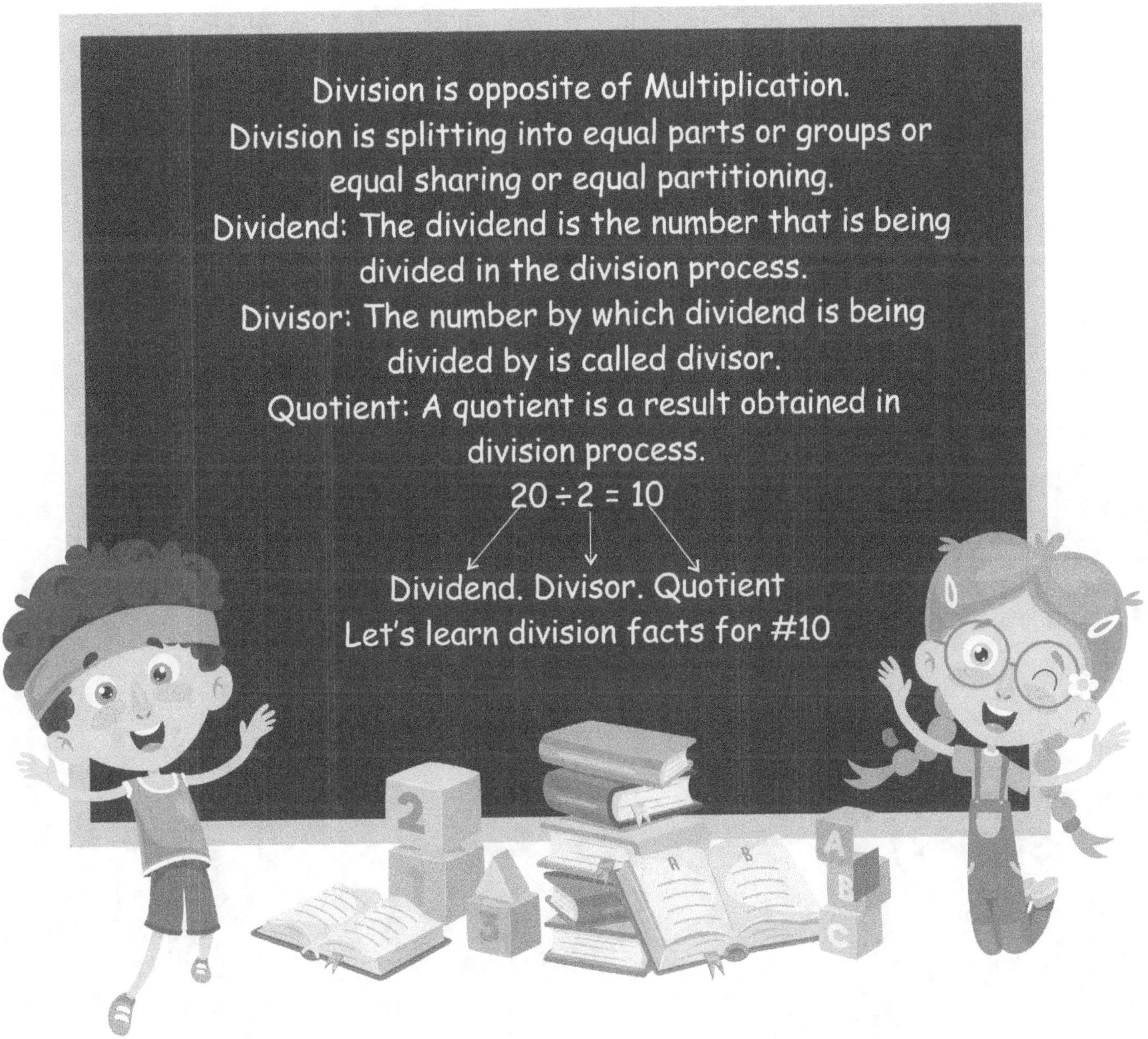

Division is opposite of Multiplication.
Division is splitting into equal parts or groups or equal sharing or equal partitioning.
Dividend: The dividend is the number that is being divided in the division process.
Divisor: The number by which dividend is being divided by is called divisor.
Quotient: A quotient is a result obtained in division process.

$$20 \div 2 = 10$$

Dividend. Divisor. Quotient
Let's learn division facts for #10

www.math-knots.com

1. Lets learn 10 ÷ 1 = 10

A.  ÷ =

B.

÷ 10

1

=

C. $10 \div 1 = 10$

www.math-knots.com

2. Lets learn $20 \div 10 = 2$

A.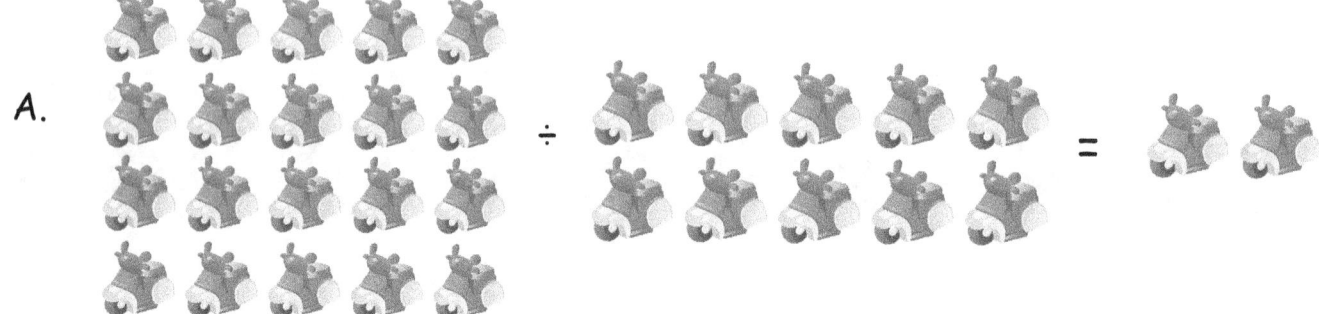

B.

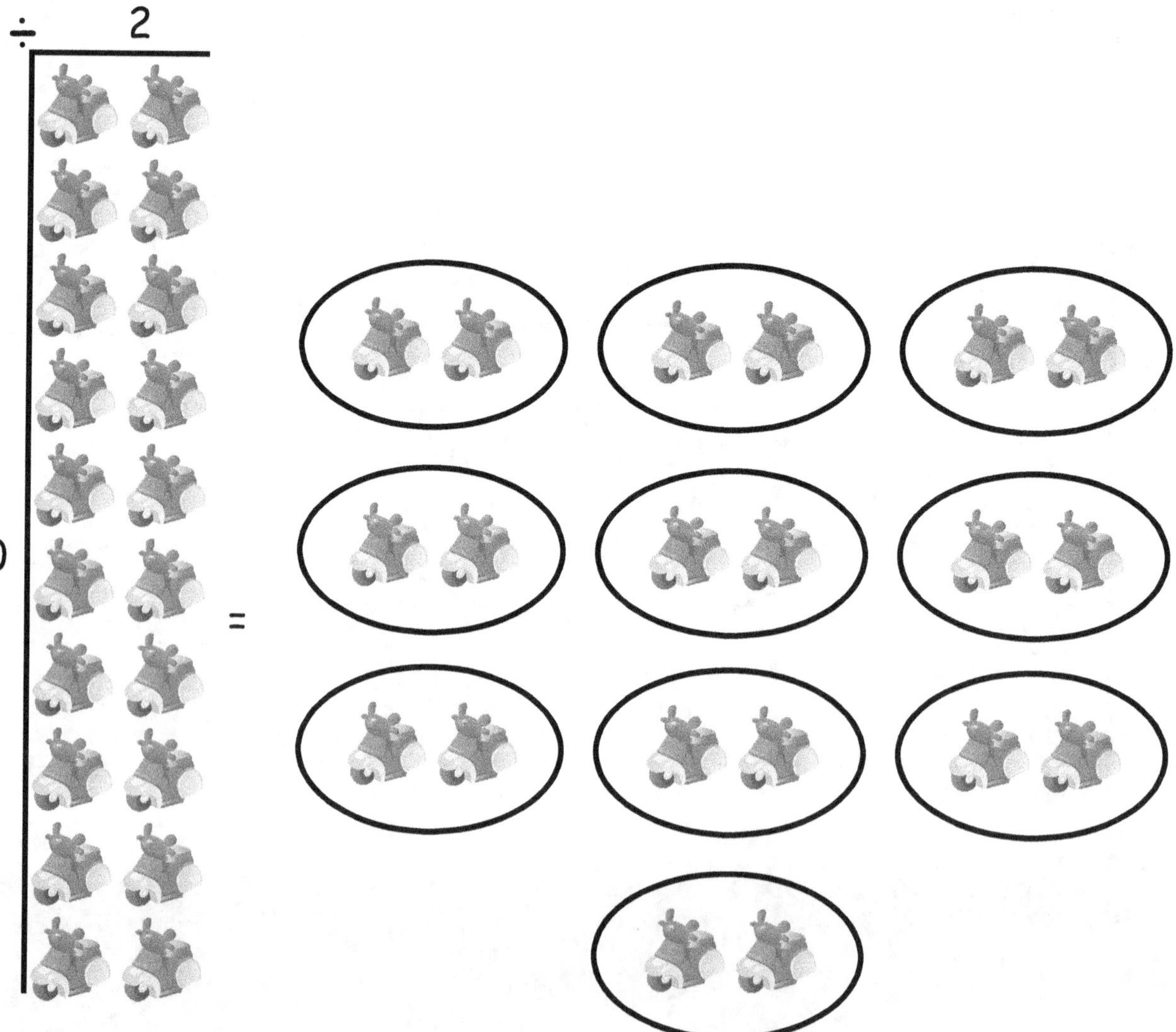

C. $20 \div 10 = 2$

3. Lets learn 30 ÷ 10 = 3

A. 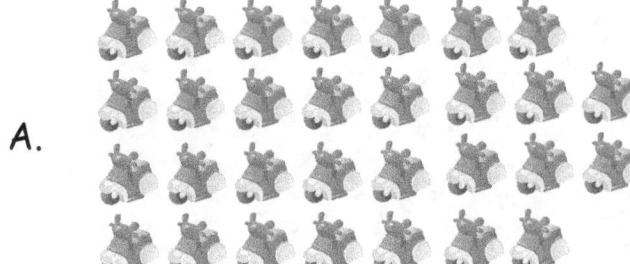 ÷  =

B.

$\div$ 3

10

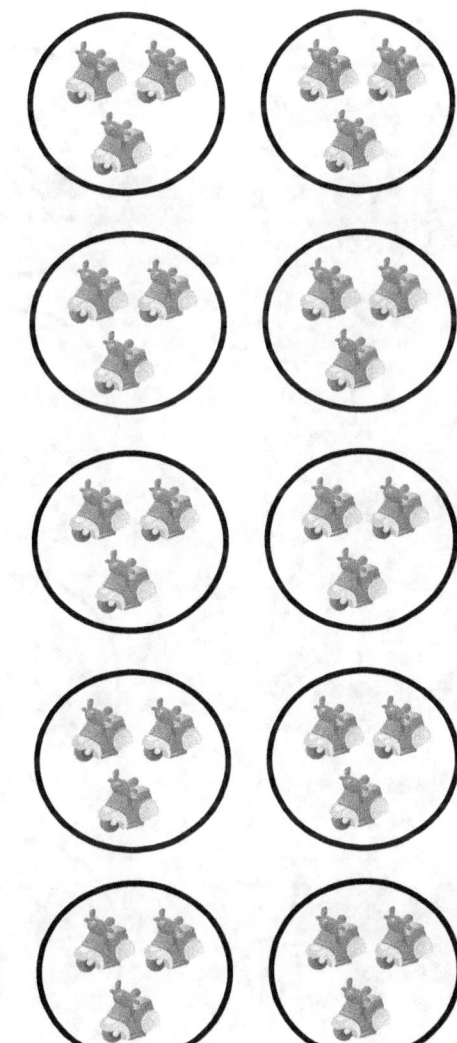

=

C. | 30 | 10 = 3 |

$\div$

4. Lets learn $40 \div 10 = 4$

A.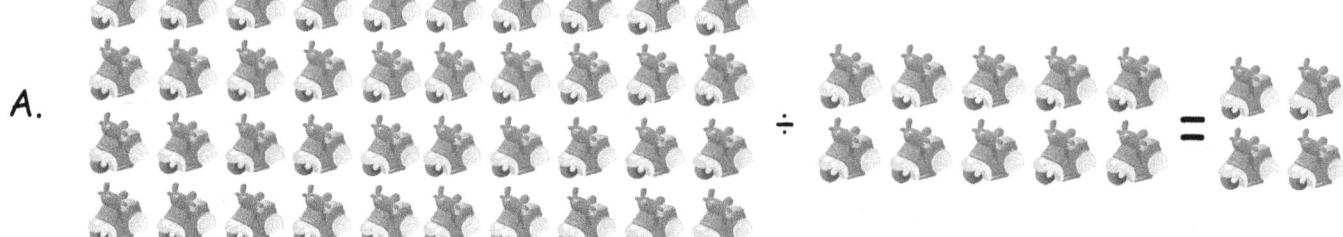

B.

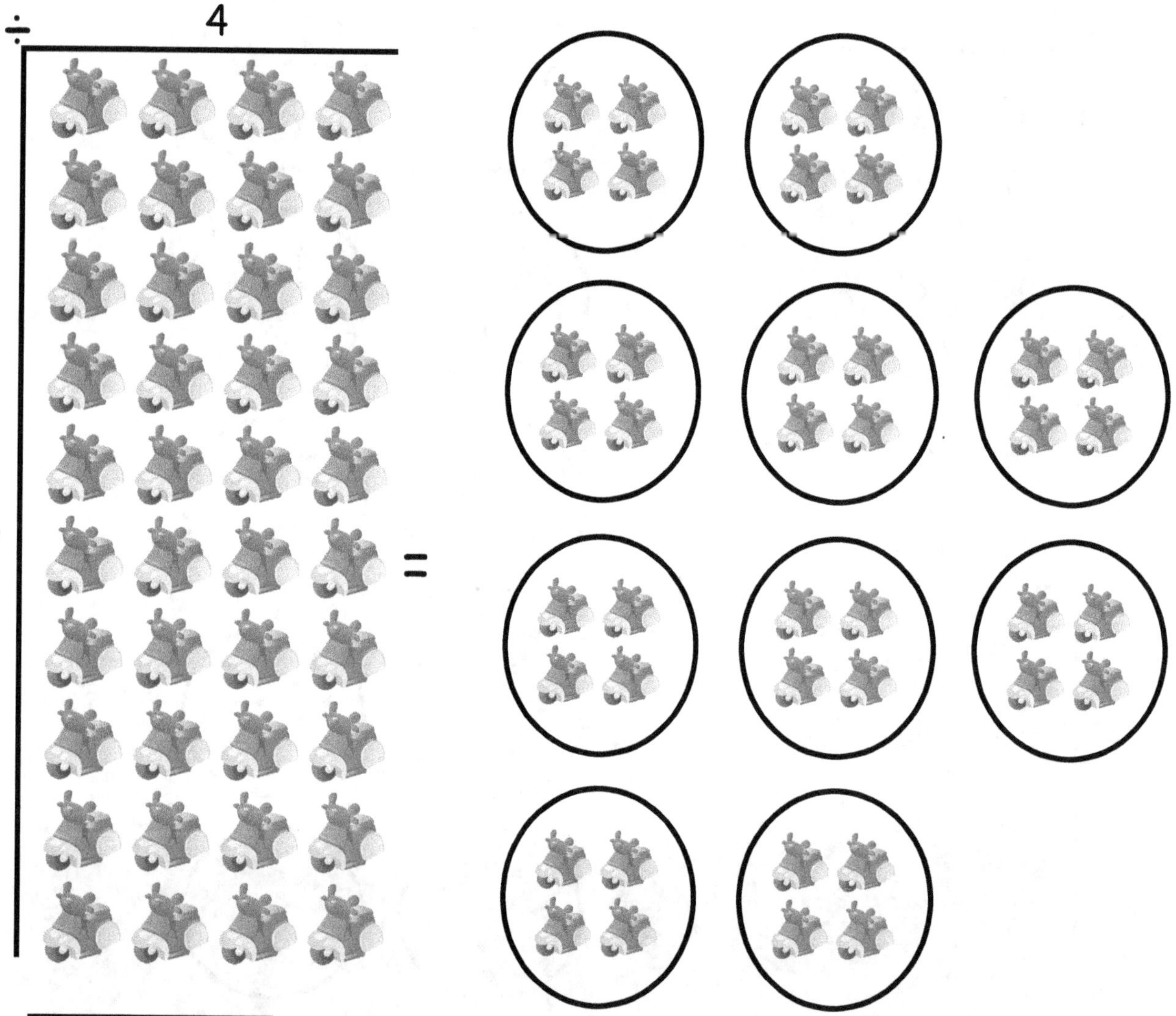

C. $\boxed{40 \div 10 = 4}$

www.math-knots.com

5. Lets learn 50 ÷ 10 = 5

A.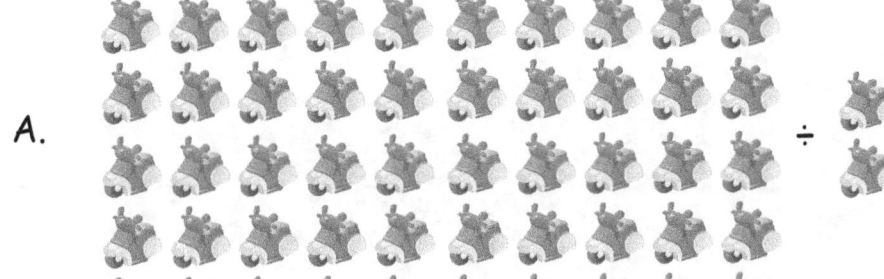

B.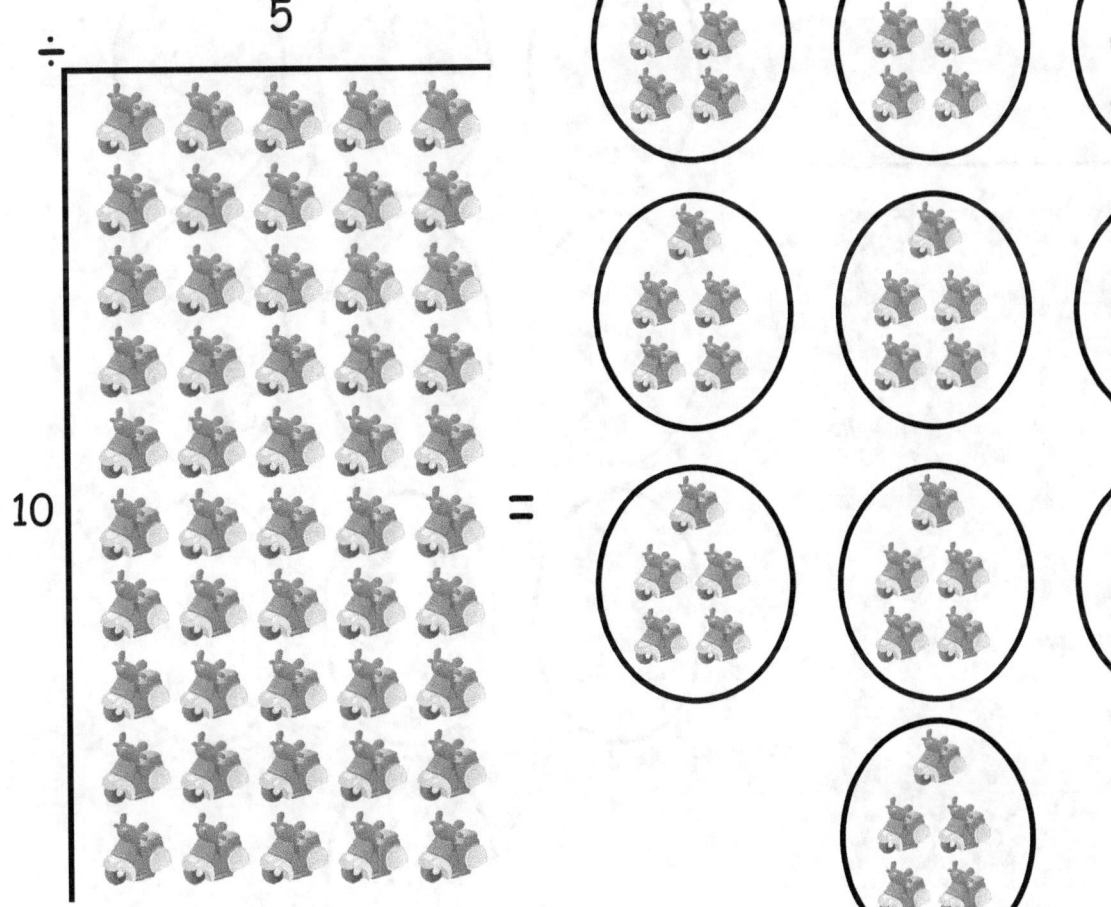

C. $50 \div 10 = 5$

6. Lets learn $60 \div 10 = 6$

A.

B.

$\div$ 6

10

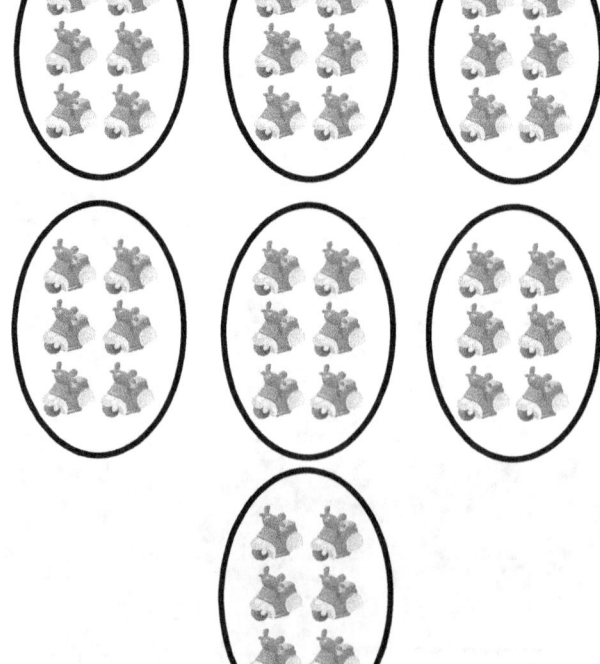

=

C. $\boxed{60 \div 10 = 6}$

7. Lets learn 70 ÷ 10 = 7

A.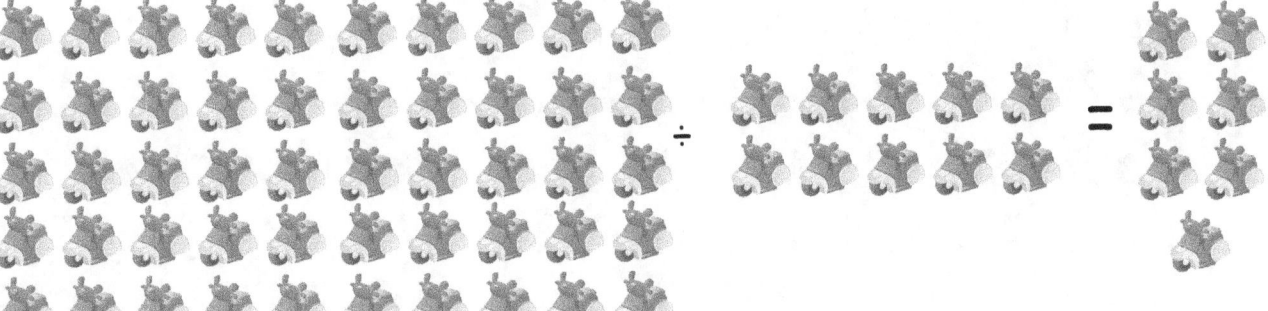

B.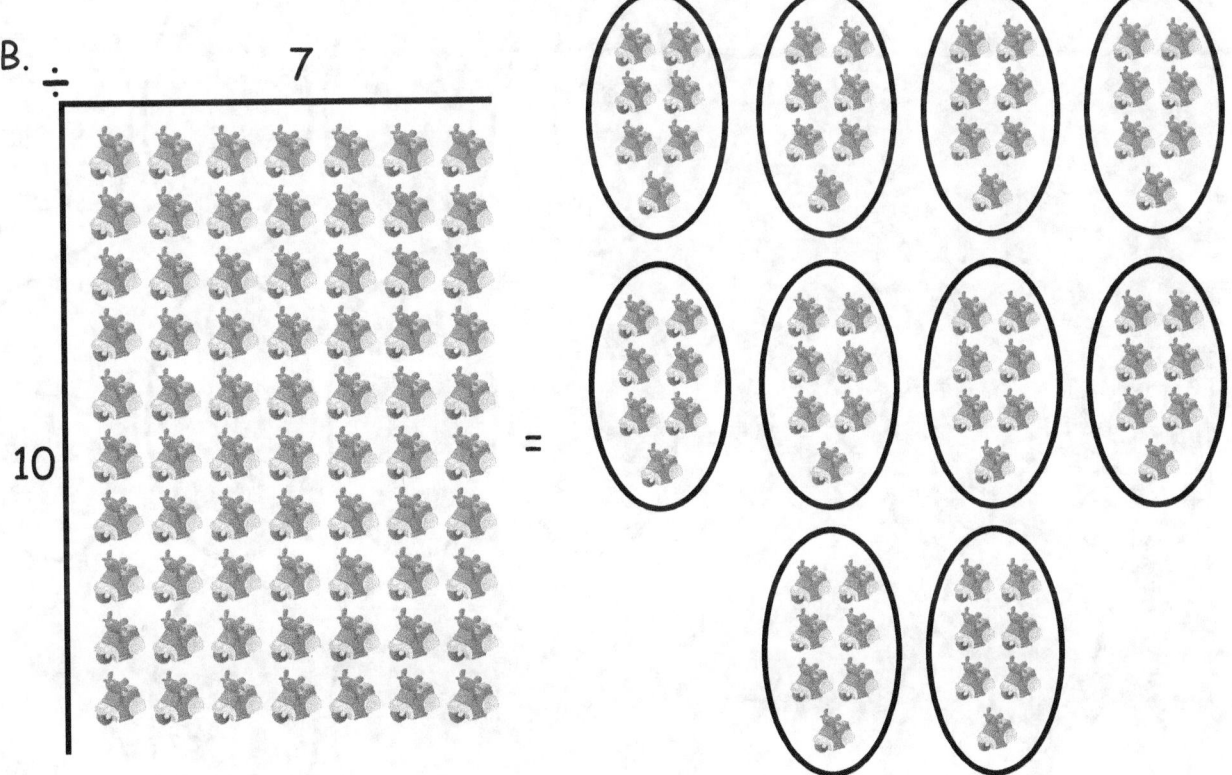

C. $70 \div 10 = 7$

8. Lets learn 80 ÷ 10 = 8

A.

 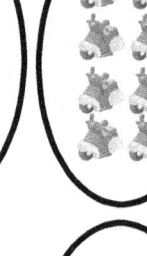 

B.

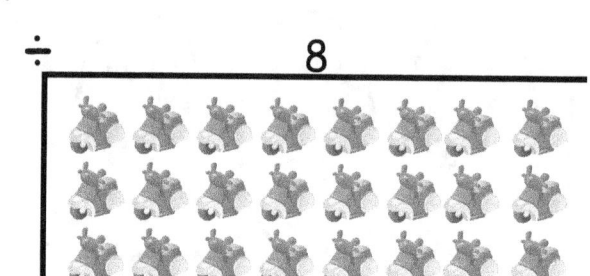

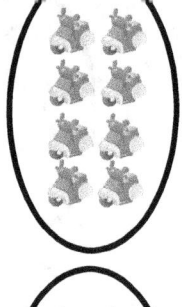

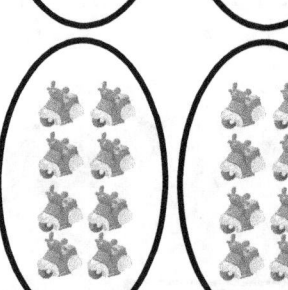

C. $\boxed{80 \div 10 = 8}$

www.math-knots.com

9.   Lets learn $90 \div 10 = 9$

A.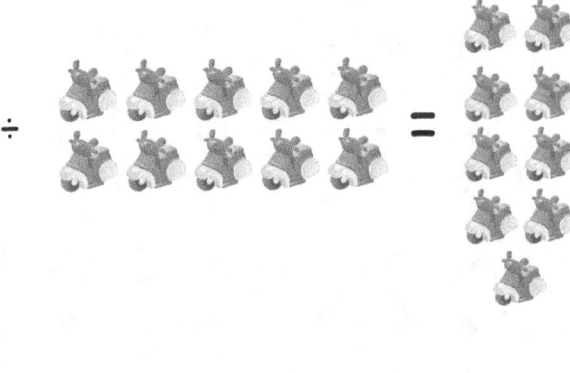

B.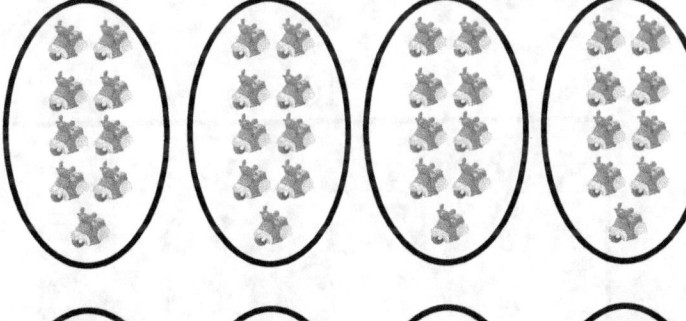

10

C. $90 \div 10 = 9$

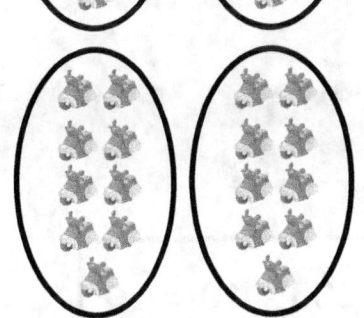

www.math-knots.com

10. Lets learn $100 \div 10 = 10$

A.

B.

$\div$

10

10

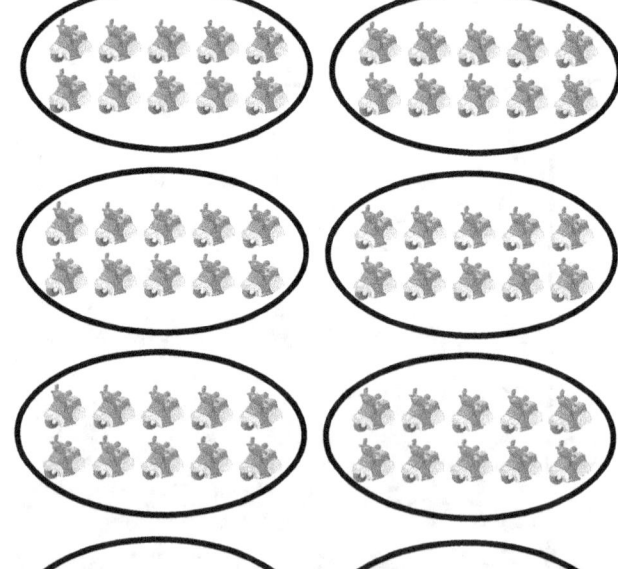

C.
$$100 \div 10 = 10$$

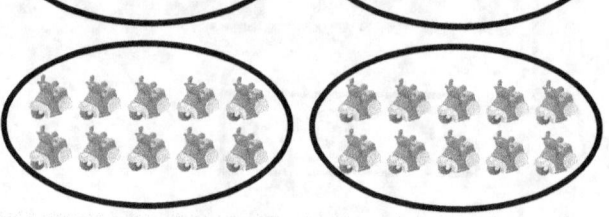

www.math-knots.com

11. Lets learn 110 ÷10 = 11

A.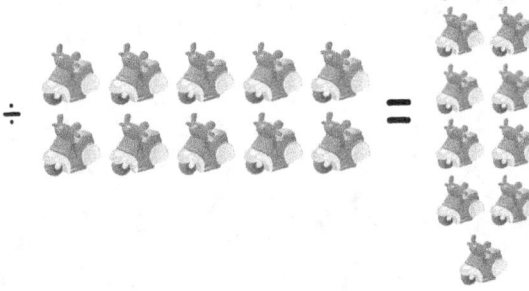

B.

$$\div \quad 11$$

10

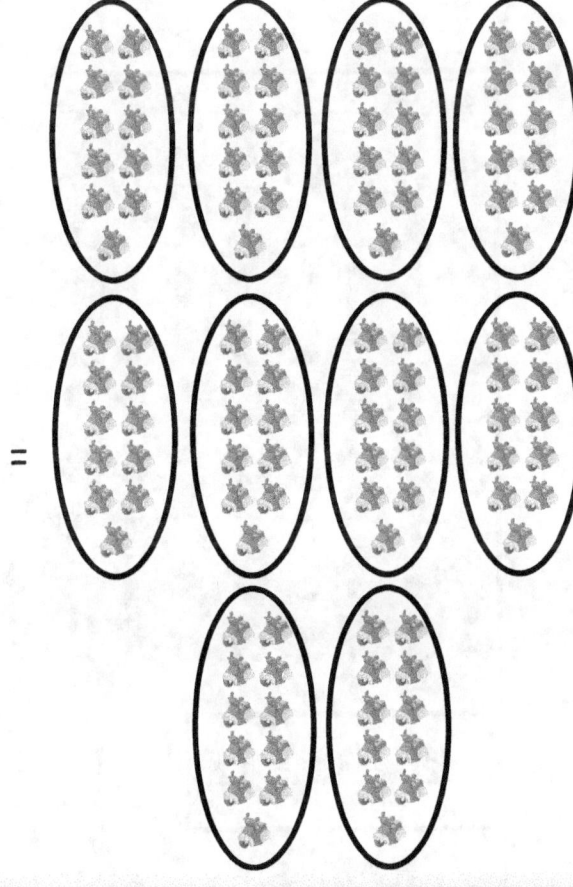

C. $\boxed{110 \div 10 = 11}$

www.math-knots.com

12.   Lets learn 120 ÷ 10 = 12

A.    ÷ 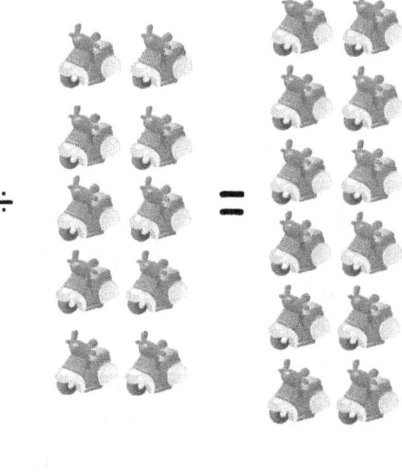 =

B.

 =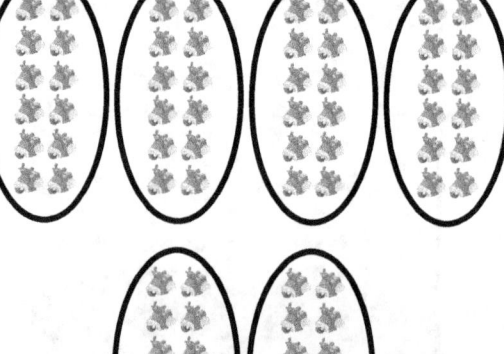

C. | 120 ÷ 10 = 12 |

www.math-knots.com

Exercise - 1

(A) 10$\overline{)10}$ (F) 10$\overline{)60}$ (K) 10$\overline{)110}$

(B) 10$\overline{)20}$ (G) 10$\overline{)70}$ (L) 10$\overline{)120}$

(C) 10$\overline{)30}$ (H) 10$\overline{)80}$ (M) 10$\overline{)130}$

(D) 10$\overline{)40}$ (I) 10$\overline{)90}$ (N) 10$\overline{)140}$

(E) 10$\overline{)50}$ (J) 10$\overline{)100}$ (O) 10$\overline{)150}$

 www.math-knots.com

Exercise - 2

1. $10 \div 10 =$ _____

2. $20 \div 10 =$ _____

3. $30 \div 10 =$ _____

4. $40 \div 10 =$ _____

5. $50 \div 10 =$ _____

6. $60 \div 10 =$ _____

7. $70 \div 10 =$ _____

8. $80 \div 10 =$ _____

9. $90 \div 10 =$ _____

10. $100 \div 10 =$ _____

11. $110 \div 10 =$ _____

12. $120 \div 10 =$ _____

1. $1 \times$ ____ $= 10$

2. $2 \times$ ____ $= 20$

3. $3 \times$ ____ $= 30$

4. $4 \times$ ____ $= 40$

5. $5 \times$ ____ $= 50$

6. $6 \times$ ____ $= 60$

7. $7 \times$ ____ $= 70$

8. $8 \times$ ____ $= 80$

9. $9 \times$ ____ $= 90$

10. $10 \times$ ____ $= 100$

11. $11 \times$ ____ $= 110$

12. $12 \times$ ____ $= 120$

Did you know division is splitting a number up by any give number.

www.math-knots.com

 ## Exercise - 3

1.   I am a number, I divide myself, into one equal group of 10. What am I ?

(A) 0   (B) 1

(C) 10   (D) 6

2.   I am a number, I divide myself, into ten equal groups of 1. What am I ?

(A) 10   (B) 50

(C) 1   (D) 2

3.   I am a number, I divide myself, into ten equal groups of 2. What am I ?

(A) 60   (B) 10

(C) 20   (D) 2

4.   I am a number, I divide myself, into ten equal groups of 3. What am I ?

(A) 10   (B) 3

(C) 18   (D) 30

5.   I am a number, I divide myself, into ten equal groups of 4. What am I ?

(A) 40   (B) 16

(C) 4   (D) 10

6.   I am a number, I divide myself, into ten equal groups of 5. What am I ?

   (A) 12   (B) 5

   (C) 10   (D) 50

7.   I am a number, I divide myself, into ten equal groups of 6. What am I ?

   (A) 36   (B) 60

   (C) 6   (D) 10

8.   I am a number, I divide myself, into ten equal groups of 7. What am I ?

   (A) 10   (B) 70

   (C) 16   (D) 7

9.   I am a number, I divide myself, into ten equal groups of 8. What am I ?

   (A) 80   (B) 8

   (C) 10   (D) 48

10. I am a number, I divide myself, into ten equal groups of 9. What am I ?

   (A) 40   (B) 9

   (C) 90   (D) 110

11.   I am a number, I divide myself, into ten equal groups of 10. What am I ?

(A)   20   (B)   60

(C)   10   (D)   100

12.   I am a number, I divide myself, into ten equal groups of 11. What am I ?

(A)   110   (B)   30

(C)   11   (D)   10

13.   I am a number, I divide myself, into ten equal groups of 12. What am I ?

(A)   32   (B)   60

(C)   70   (D)   120

14.   I am a number, I divide myself, into ten equal groups of 13. What am I ?

(A)   10   (B)   40

(C)   130   (D)   13

15.   I am a number, I divide myself, into ten equal groups of 14. What am I ?

(A)   10   (B)   52

(C)   140   (D)   70

189

Exercise - 4

Solve the maze run below.

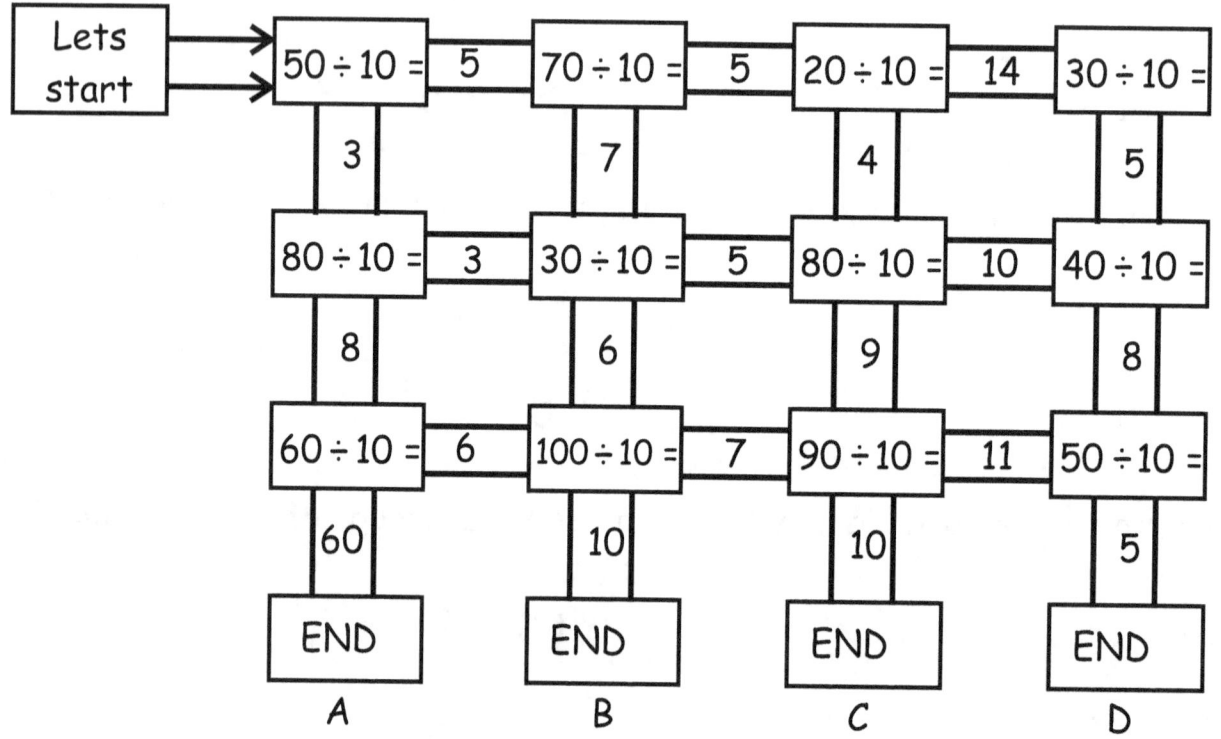

Lets start	→ →	50 ÷ 10 =	5	70 ÷ 10 =	5	20 ÷ 10 =	14	30 ÷ 10 =

Column values (vertical): 3 | 7 | 4 | 5

| 80 ÷ 10 = | 3 | 30 ÷ 10 = | 5 | 80 ÷ 10 = | 10 | 40 ÷ 10 = |

Column values (vertical): 8 | 6 | 9 | 8

| 60 ÷ 10 = | 6 | 100 ÷ 10 = | 7 | 90 ÷ 10 = | 11 | 50 ÷ 10 = |

Column values (vertical): 60 | 10 | 10 | 5

| END | END | END | END |
| A | B | C | D |

Who won the race ? _____

www.math-knots.com

Exercise - 5

1.   10 ÷ ☐ = 1 then ☐ = _____

2.   20 ÷ ☐ = 10 then ☐ = _____

3.   30 ÷ ☐ = 10 then ☐ = _____

4.   40 ÷ ☐ = 10 then ☐ = _____

5.   50 ÷ ☐ = 10 then ☐ = _____

6.   60 ÷ ☐ = 10 then ☐ = _____

7.   70 ÷ ☐ = 10 then ☐ = _____

8.   80 ÷ ☐ = 10 then ☐ = _____

9.   90 ÷ ☐ = 10 then ☐ = _____

10. 100 ÷ ☐ = 10 then ☐ = _____

11. 110 ÷ ☐ = 10 then ☐ = _____

12. 120 ÷ ☐ = 10 then ☐ = _____

Hey you are an expert of division facts #10 !!!

Division is opposite of Multiplication.
Division is splitting into equal parts or groups or equal sharing or equal partitioning.
Dividend: The dividend is the number that is being divided in the division process.
Divisor: The number by which dividend is being divided by is called divisor.
Quotient: A quotient is a result obtained in division process.

$$22 \div 11 = 2$$

Dividend. Divisor. Quotient
Let's learn division facts for #11

www.math-knots.com

1. Lets learn $11 \div 1 = 11$

A.

B.

$$\div 11$$

1

=

C. $\boxed{11 \div 1 = 11}$

www.math-knots.com

2.   Lets learn 22 ÷ 11 = 2

A.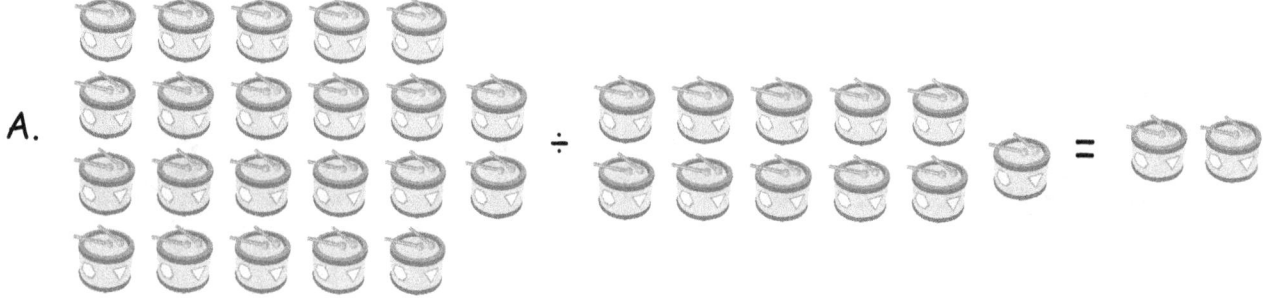

B.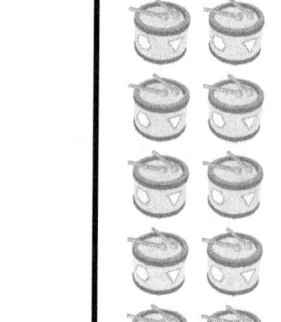

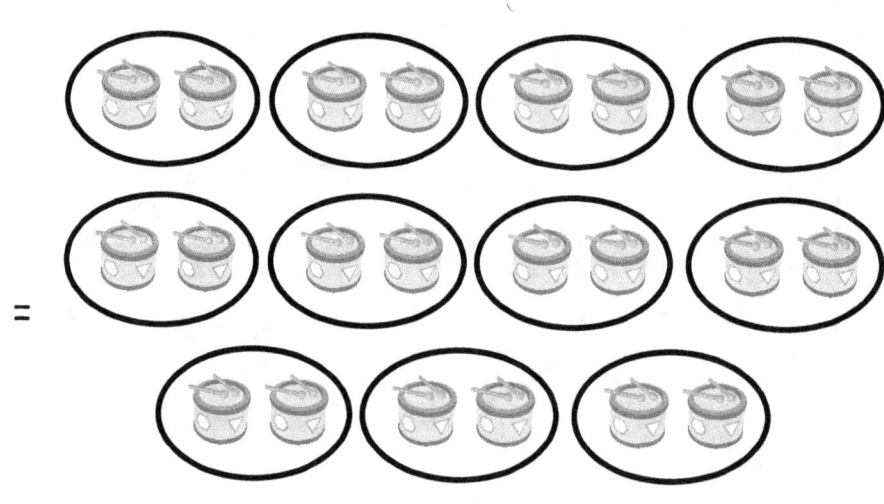

C. **22 ÷ 11 = 2**

www.math-knots.com

3. Lets learn 33 ÷ 11 = 3

A.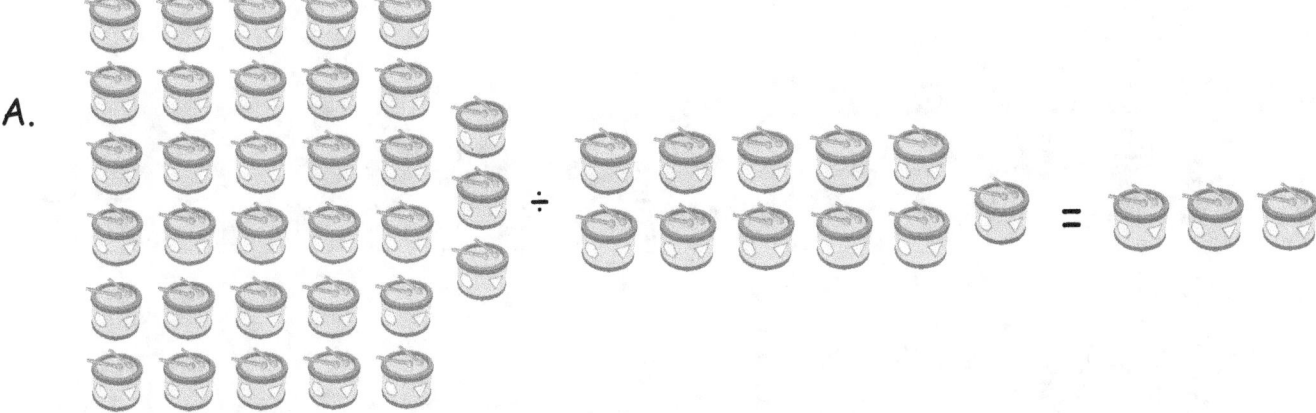

B.

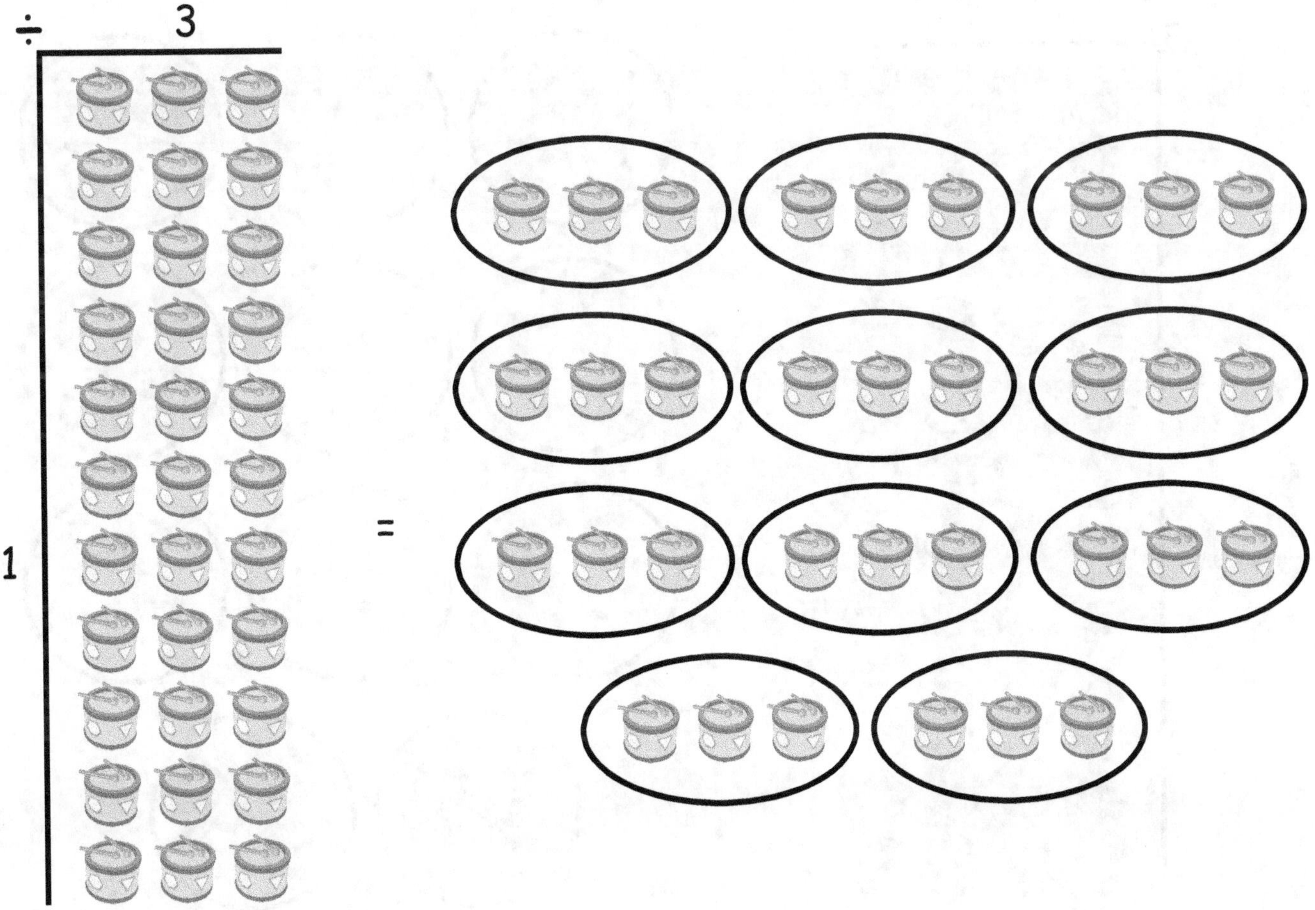

C. | 33 ÷ 11 = 3 |

4. Lets learn 44 ÷ 11 = 4

A.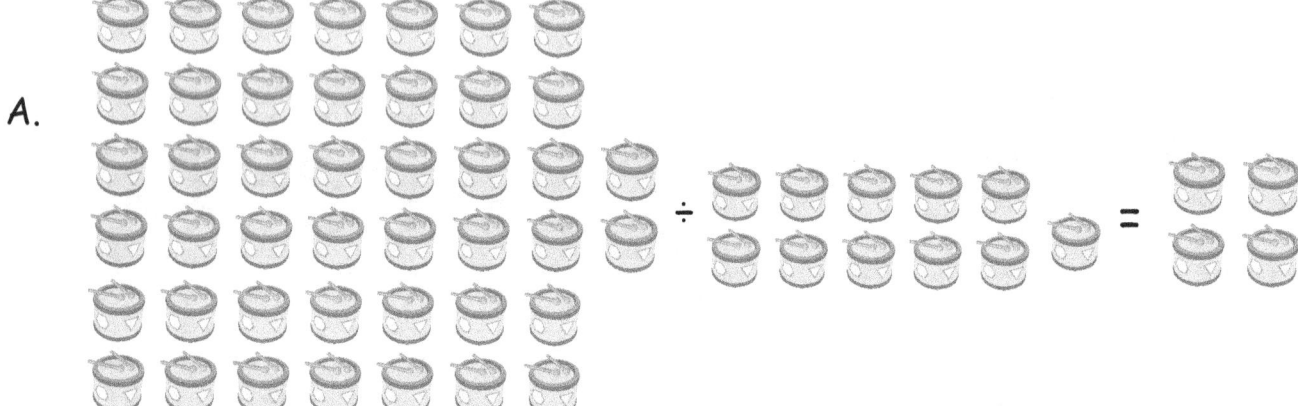

B.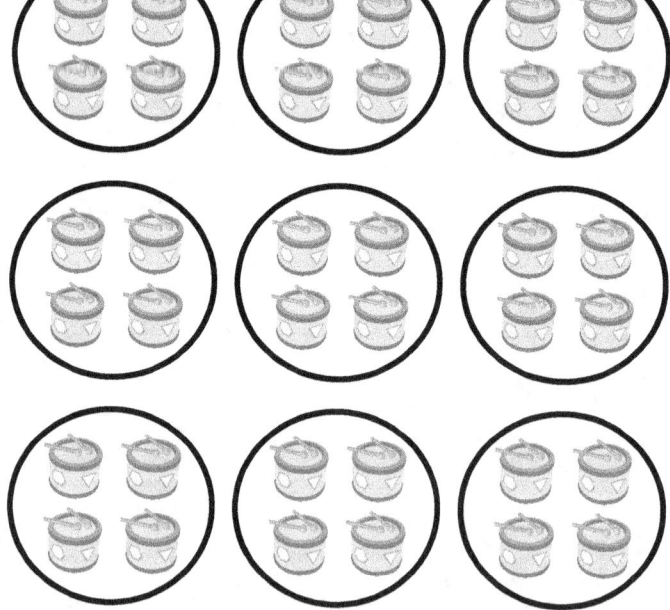

C. $44 \div 11 = 4$

5. Lets learn 55 ÷ 11 = 5

A.

B. ÷   5

11 =

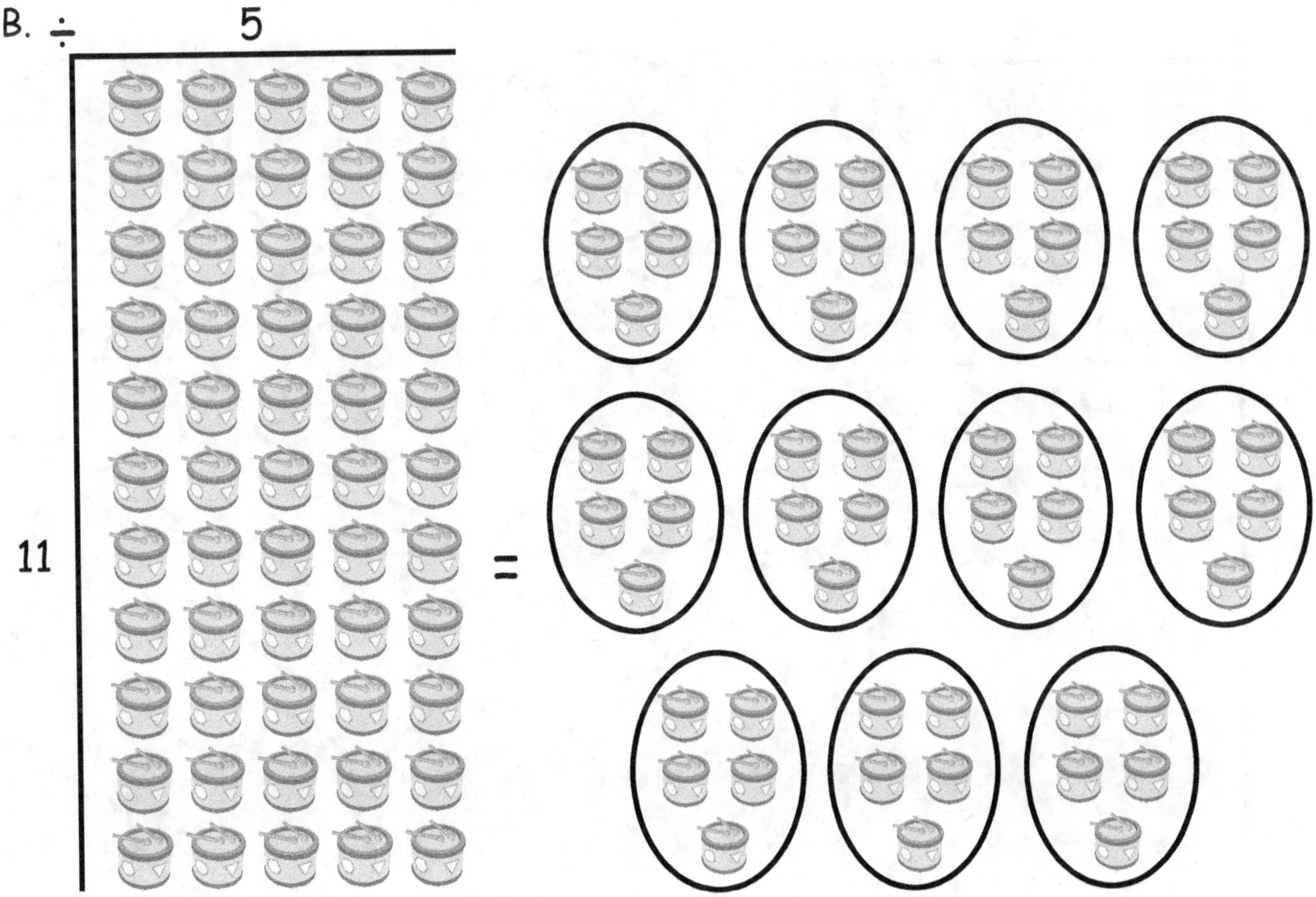

C. **55 ÷ 11 = 5**

6.   Lets learn 66 ÷ 11 = 6

A.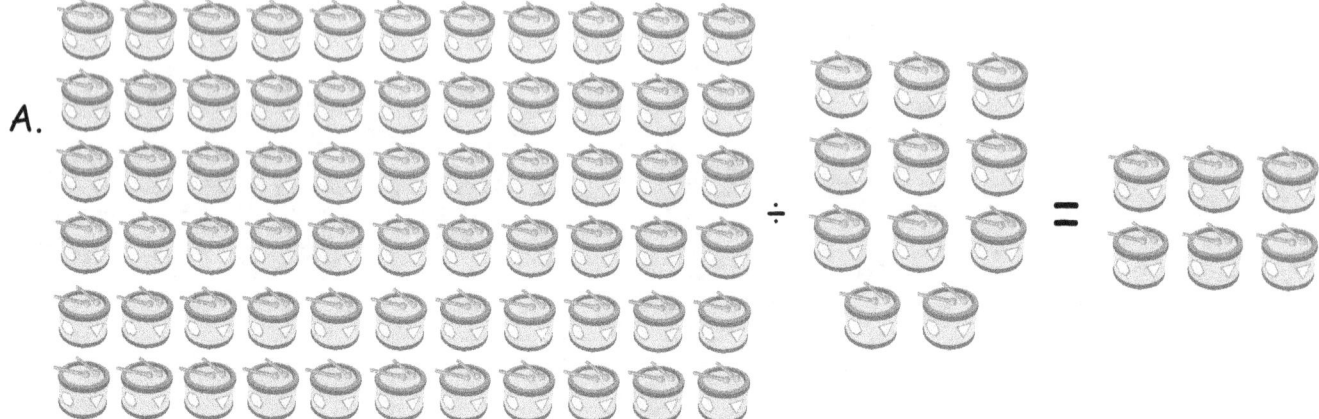

B. ÷

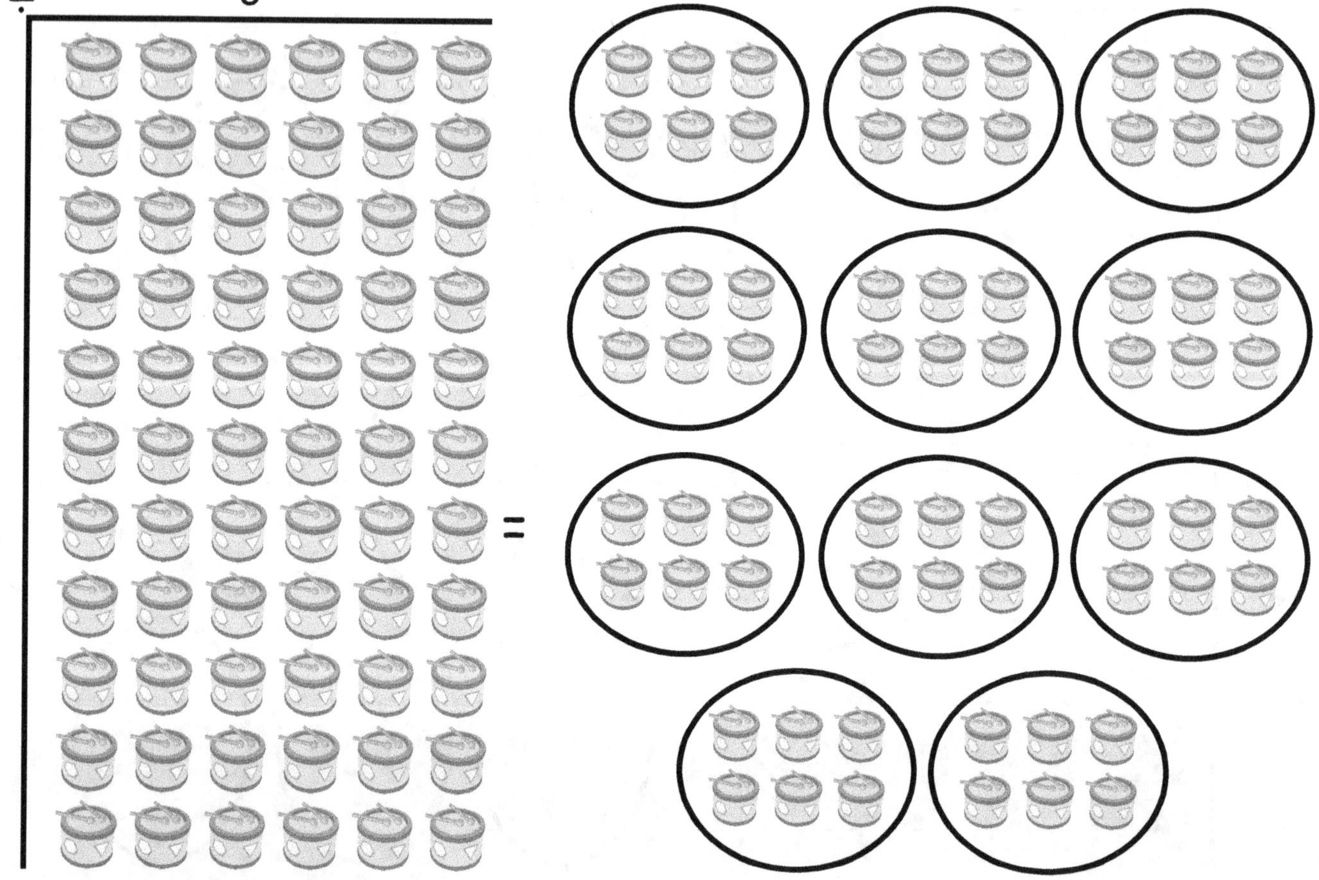

C. $$66 \div 11 = 6$$

www.math-knots.com

7. Lets learn 77 ÷ 11 = 7

A.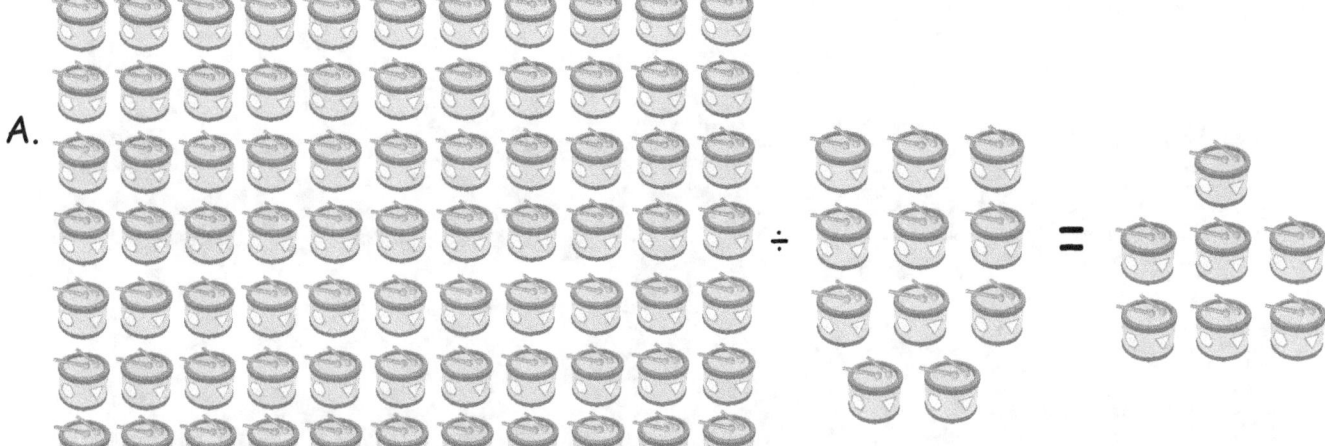

B.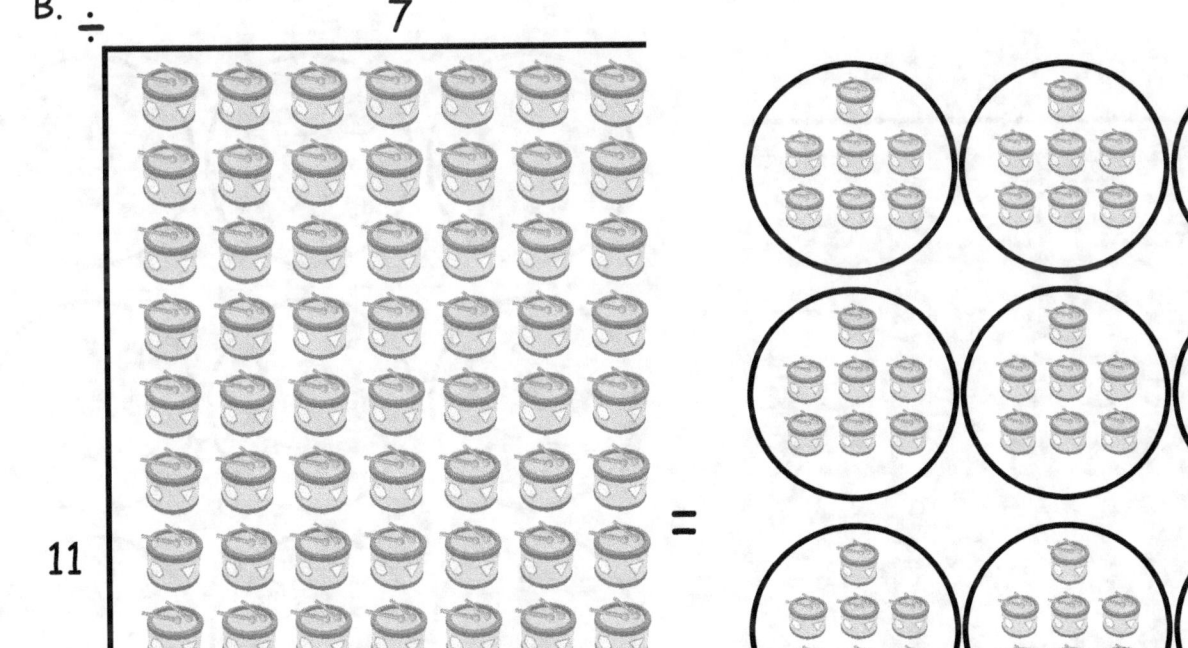

C. 77 ÷ 11 = 7

www.math-knots.com

8. Lets learn 88 ÷ 11 = 8

A.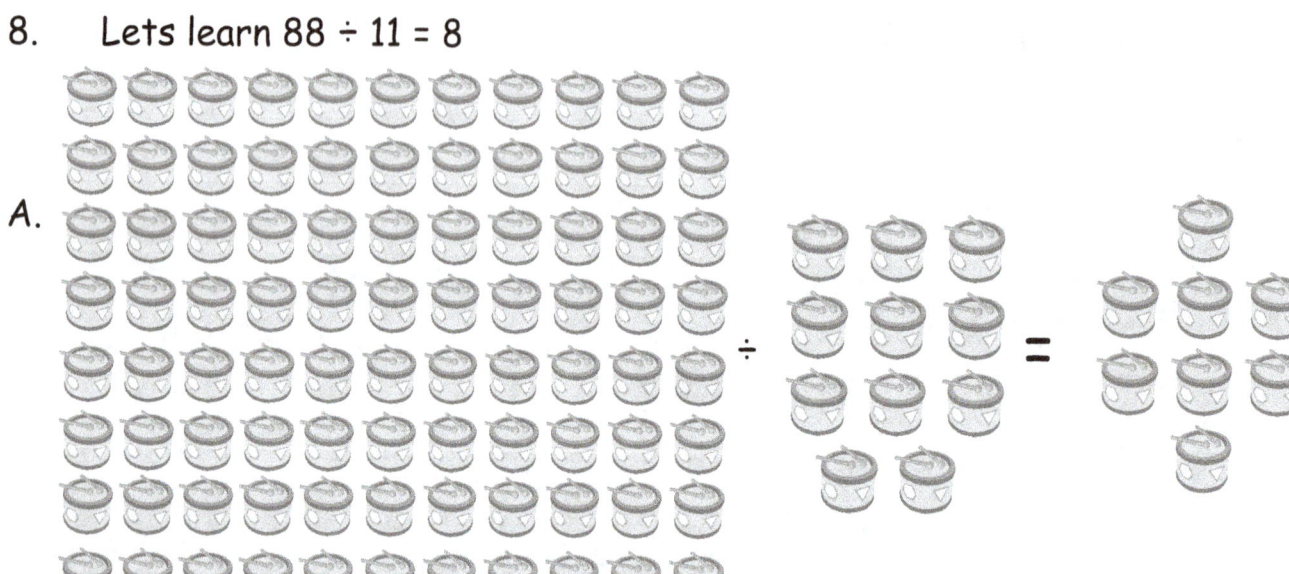

B. ÷

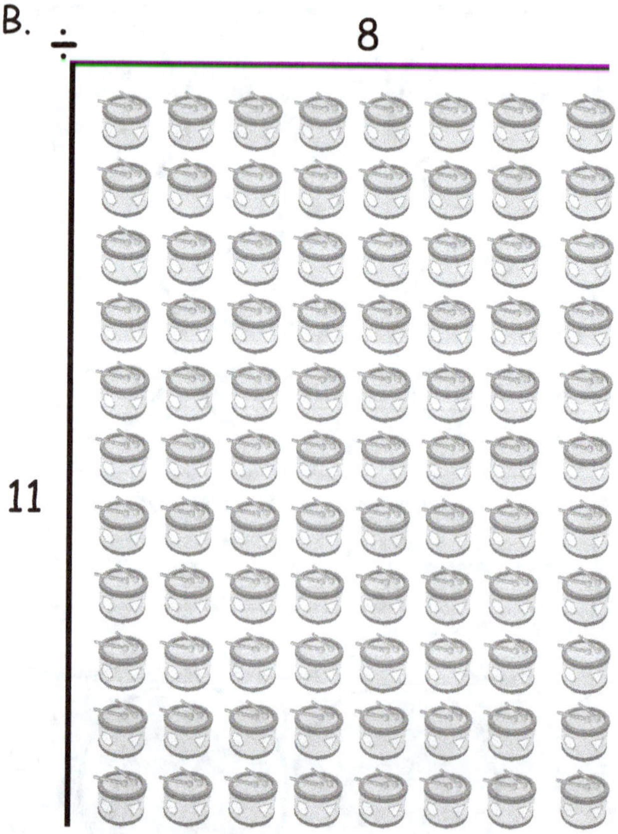

8

11

=

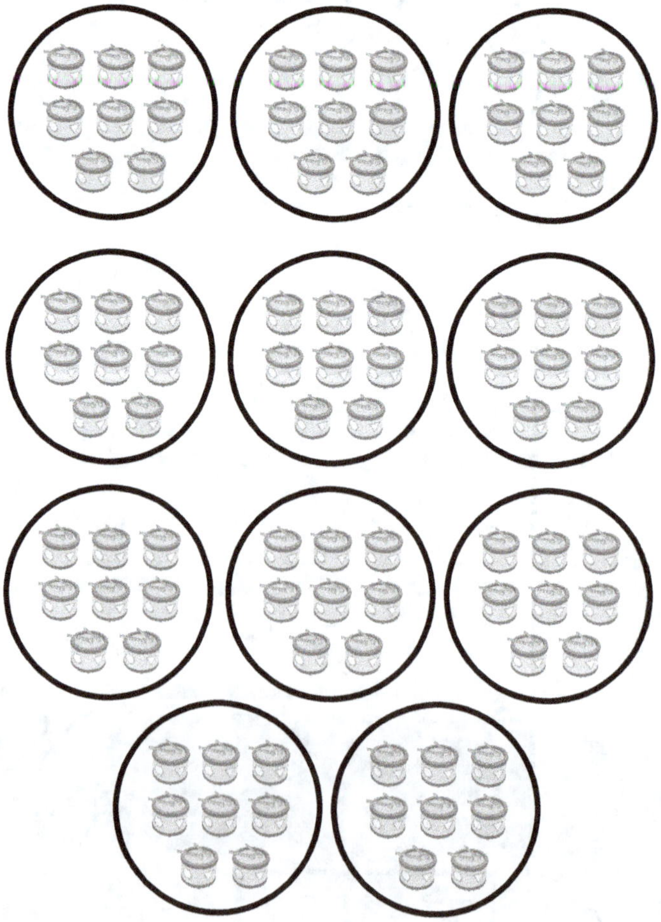

C.

88 ÷ 11 = 8

9.   Lets learn 99 ÷ 11 = 9

A.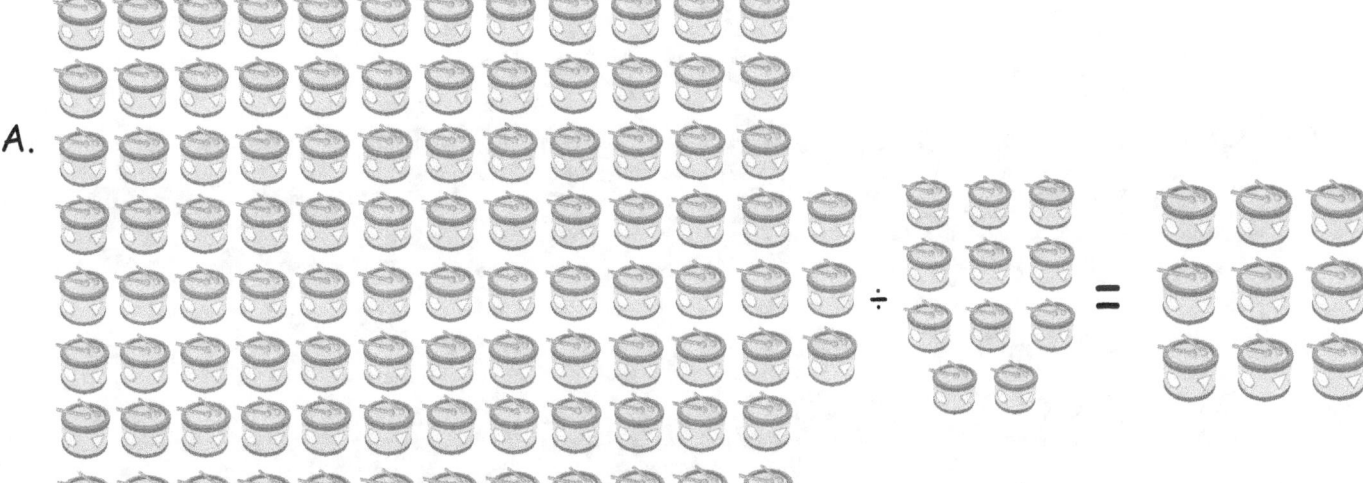

B.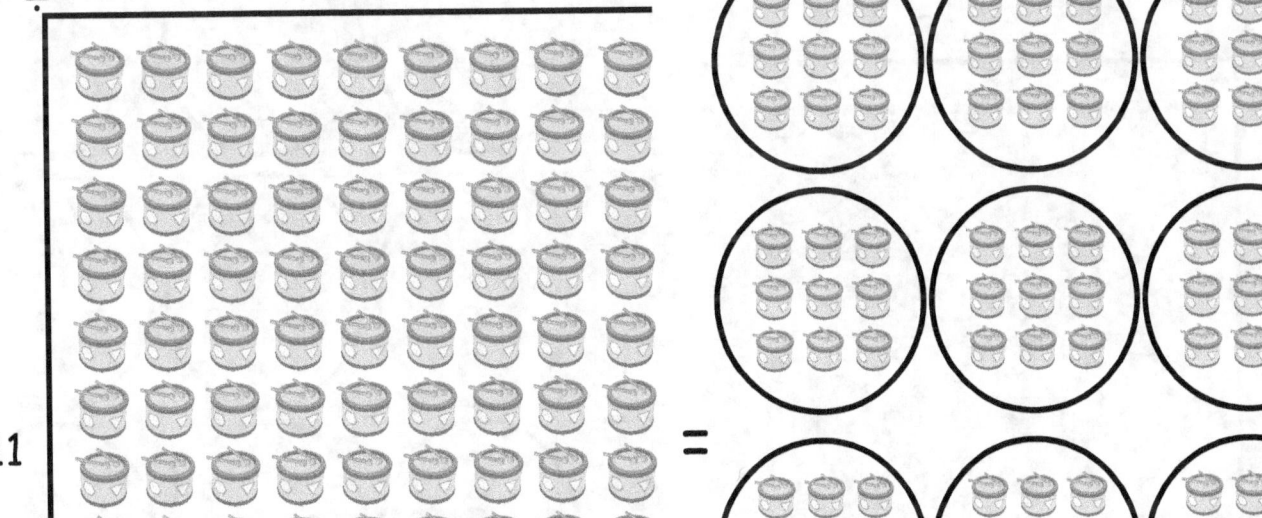

÷   9

11

=

C.   **99 ÷ 11 = 9**

www.math-knots.com

10. Lets learn 110 ÷ 11 = 10

A.

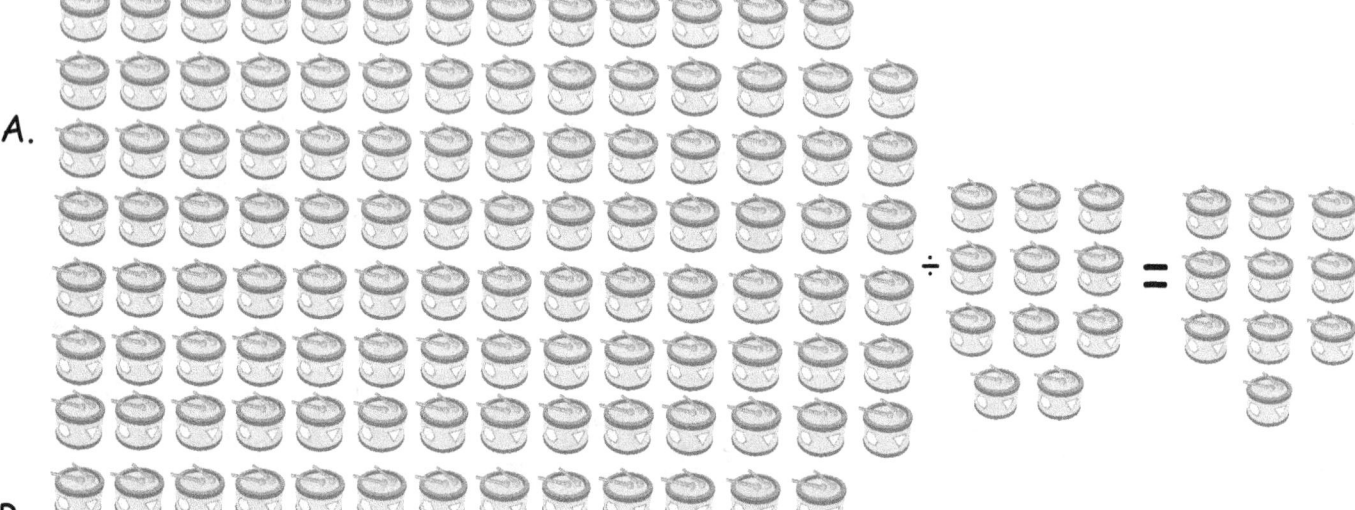

B.

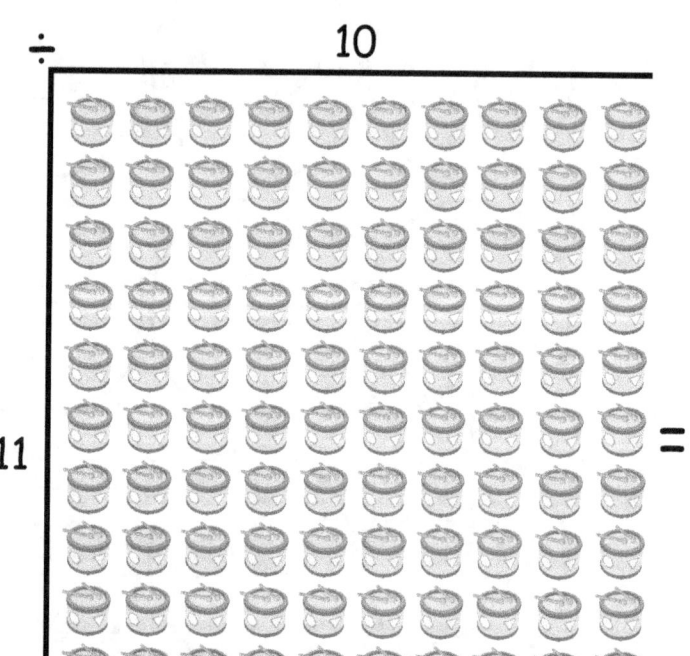

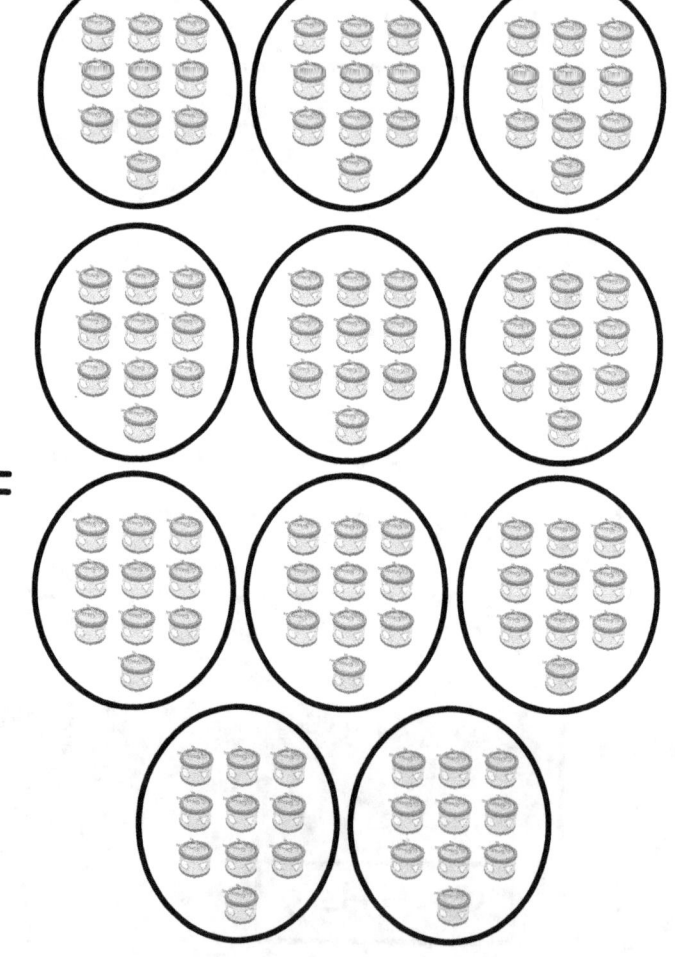

C. 110 ÷ 11 = 10

www.math-knots.com

11.   Lets learn 121 ÷ 11 = 11

A.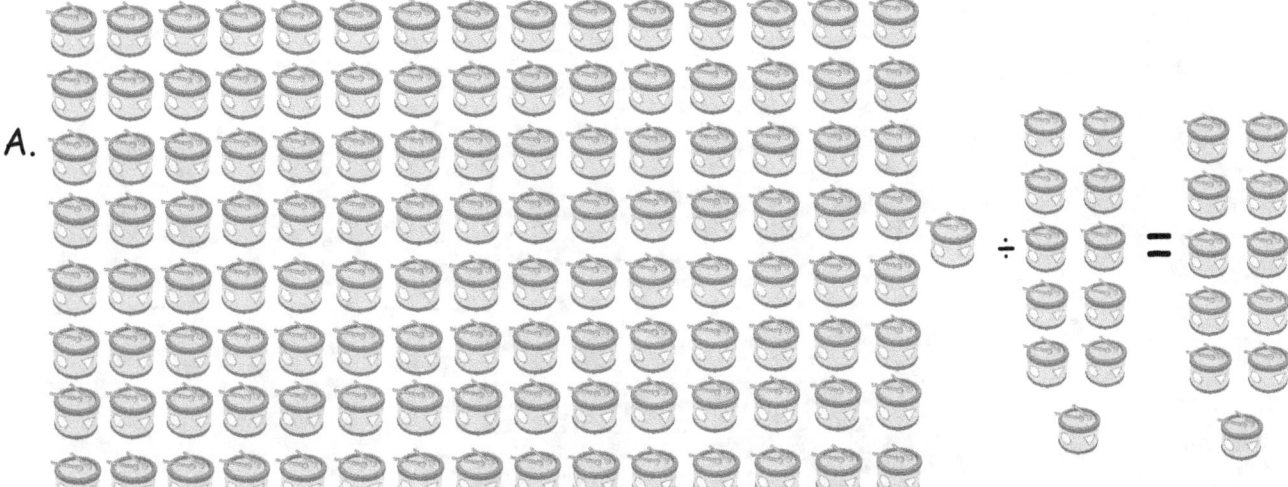

B.

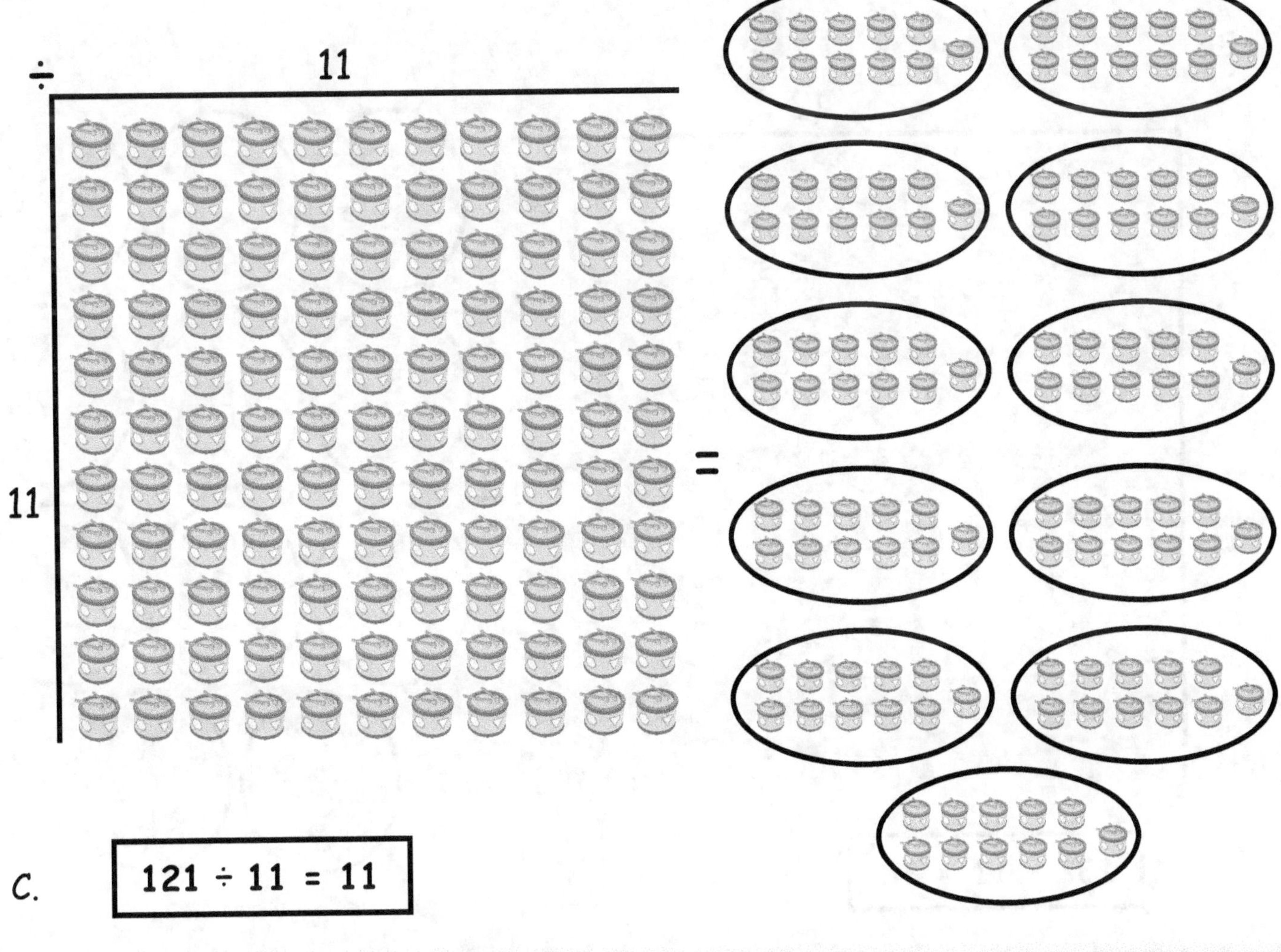

C.   **121 ÷ 11 = 11**

www.math-knots.com

12. Lets learn 132 ÷ 11 = 12

A.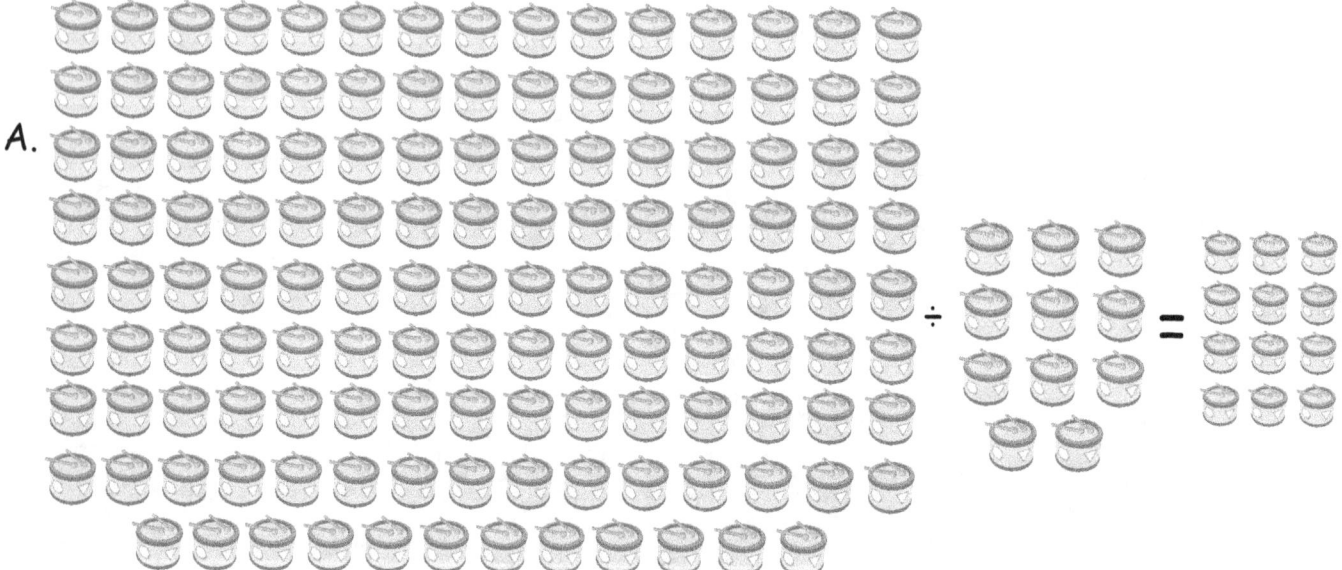

B.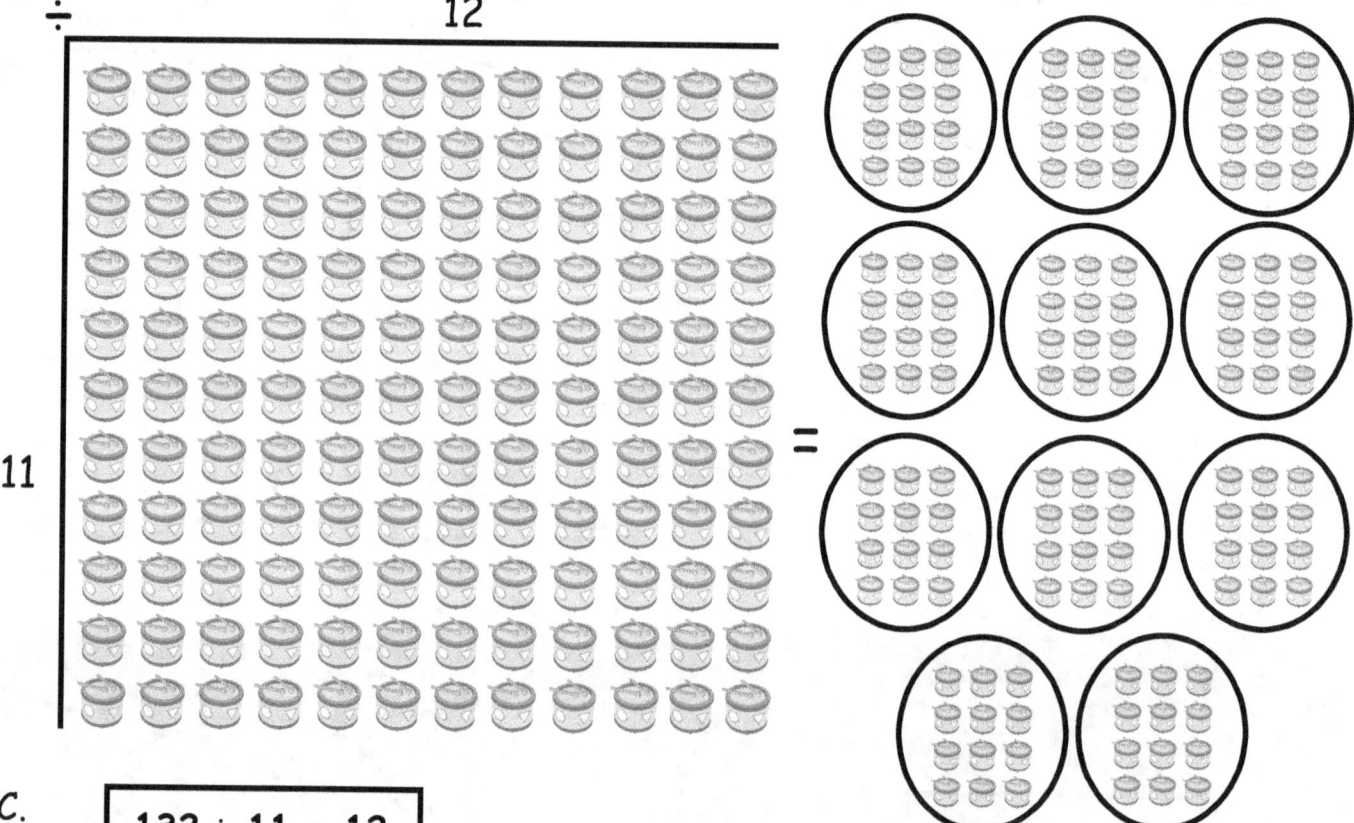

C. | 132 ÷ 11 = 12 |

www.math-knots.com

Exercise - 1

(A) $11\overline{)11}$ (F) $11\overline{)66}$ (K) $11\overline{)121}$

(B) $11\overline{)22}$ (G) $11\overline{)77}$ (L) $11\overline{)132}$

(C) $11\overline{)33}$ (H) $11\overline{)88}$ (M) $11\overline{)143}$

(D) $11\overline{)44}$ (I) $11\overline{)99}$ (N) $11\overline{)154}$

(E) $11\overline{)55}$ (J) $11\overline{)110}$ (O) $11\overline{)165}$

www.math-knots.com

Exercise - 2

1. 11 ÷ 11 = _____

2. 22 ÷ 11 = _____

3. 33 ÷ 11 = _____

4. 44 ÷ 11 = _____

5. 55 ÷ 11 = _____

6. 66 ÷ 11 = _____

7. 77 ÷ 11 = _____

8. 88 ÷ 11 = _____

9. 99 ÷ 11 = _____

10. 110 ÷ 11 = _____

11. 121 ÷ 11 = _____

12. 132 ÷ 11 = _____

1 × ___ = 11

2 × ___ = 22

3 × ___ = 33

4 × ___ = 44

5 × ___ = 55

6 × ___ = 66

7 × ___ = 77

8 × ___ = 88

9 × ___ = 99

10 × ___ = 110

11 × ___ = 121

12 × ___ = 132

Did you know division is splitting a number up by any give number.

www.math-knots.com

Exercise - 3

1. I am a number, I divide myself, into one equal group of 11. What am I ?

   (A) 0   (B) 11

   (C) 10   (D) 6

2. I am a number, I divide myself, into eleven equal groups of 1. What am I ?

   (A) 11   (B) 6

   (C) 6   (D) 2

3. I am a number, I divide myself, into eleven equal groups of 2. What am I ?

   (A) 6   (B) 12

   (C) 22   (D) 4

4. I am a number, I divide myself, into eleven equal groups of 3. What am I ?

   (A) 33   (B) 4

   (C) 18   (D) 2

5. I am a number, I divide myself, into eleven equal groups of 4. What am I ?

   (A) 24   (B) 16

   (C) 8   (D) 44

www.math-knots.com

6. I am a number, I divide myself, into eleven equal groups of 5. What am I ?

    (A) 55   (B) 5

    (C) 30   (D) 11

7. I am a number, I divide myself, into eleven equal groups of 6. What am I ?

    (A) 36   (B) 6

    (C) 11   (D) 66

8. I am a number, I divide myself, into eleven equal groups of 7. What am I ?

    (A) 11   (B) 7

    (C) 77   (D) 44

9. I am a number, I divide myself, into eleven equal groups of 8. What am I ?

    (A) 11   (B) 8

    (C) 22   (D) 88

10. I am a number, I divide myself, into eleven equal groups of 9. What am I ?

    (A) 36   (B) 99

    (C) 9   (D) 54

www.math-knots.com

11. I am a number, I divide myself, into eleven equal groups of 10. What am I ?

(A) 110   (B) 66

(C) 22   (D) 11

12. I am a number, I divide myself, into eleven equal groups of 11. What am I ?

(A) 55   (B) 44

(C) 22   (D) 121

13. I am a number, I divide myself, into eleven equal groups of 12. What am I ?

(A) 132   (B) 11

(C) 12   (D) 66

14. I am a number, I divide myself, into eleven equal groups of 13. What am I ?

(A) 36   (B) 13

(C) 143   (D) 77

15. I am a number, I divide myself, into eleven equal groups of 14. What am I ?

(A) 14   (B) 55

(C) 88   (D) 154

Exercise - 4

Solve the maze run below.

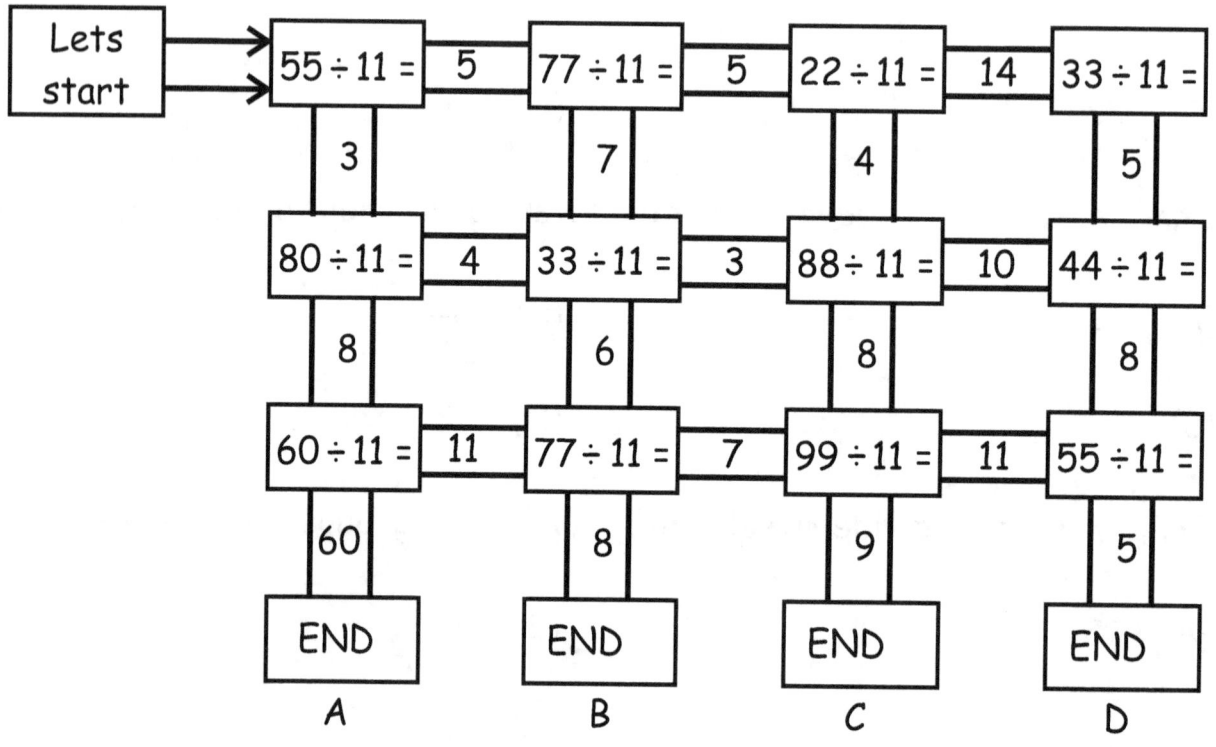

Lets start	→	$55 \div 11 =$	5	$77 \div 11 =$	5	$22 \div 11 =$	14	$33 \div 11 =$
		3		7		4		5
		$80 \div 11 =$	4	$33 \div 11 =$	3	$88 \div 11 =$	10	$44 \div 11 =$
		8		6		8		8
		$60 \div 11 =$	11	$77 \div 11 =$	7	$99 \div 11 =$	11	$55 \div 11 =$
		60		8		9		5
		END A		END B		END C		END D

Who won the race ? _____

www.math-knots.com

Exercise - 5

1. $11 \div \square = 1$ then $\square$ = _____

2. $22 \div \square = 11$ then $\square$ = _____

3. $33 \div \square = 11$ then $\square$ = _____

4. $44 \div \square = 11$ then $\square$ = _____

5. $55 \div \square = 11$ then $\square$ = _____

6. $66 \div \square = 11$ then $\square$ = _____

7. $77 \div \square = 11$ then $\square$ = _____

8. $88 \div \square = 11$ then $\square$ = _____

9. $99 \div \square = 11$ then $\square$ = _____

10. $110 \div \square = 11$ then $\square$ = _____

11. $121 \div \square = 11$ then $\square$ = _____

12. $132 \div \square = 11$ then $\square$ = _____

Hey you are an expert of division facts of #11 !!!

www.math-knots.com

Division is opposite of Multiplication.
Division is splitting into equal parts or groups or equal sharing or equal partitioning.
Dividend: The dividend is the number that is being divided in the division process.
Divisor: The number by which dividend is being divided by is called divisor.
Quotient: A quotient is a result obtained in division process.

$$24 \div 12 = 2$$

Dividend. Divisor. Quotient
Let's learn division facts for #12

www.math-knots.com

1. Lets learn 12 ÷ 1 = 12

A. ⬤⬤⬤⬤⬤⬤ ÷ ⬤ =
   ⬤⬤⬤⬤⬤⬤

B.
÷ 12

1 =

C. **12 ÷ 1 = 12**

2. Lets learn 24 ÷ 12 = 2

A.

B.

$$\div \quad 2$$

12

=

C. | 24 ÷ 12 = 2 |

3. Lets learn $36 \div 12 = 3$

A.

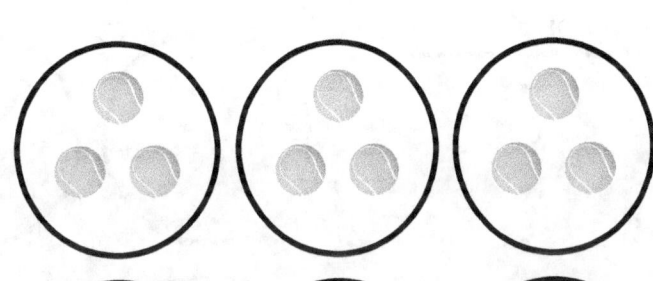

B.

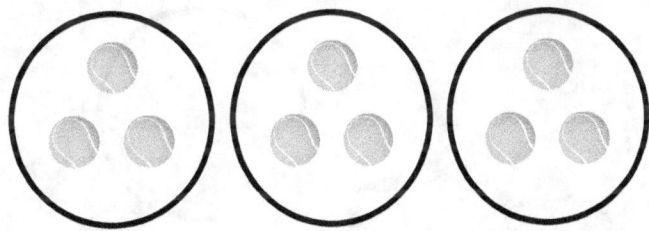

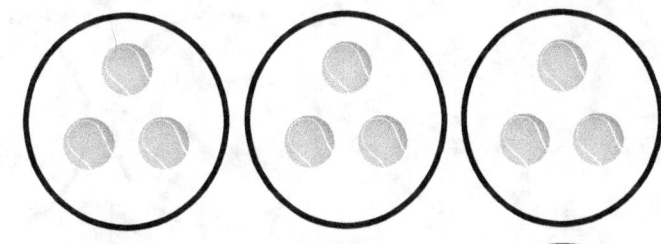

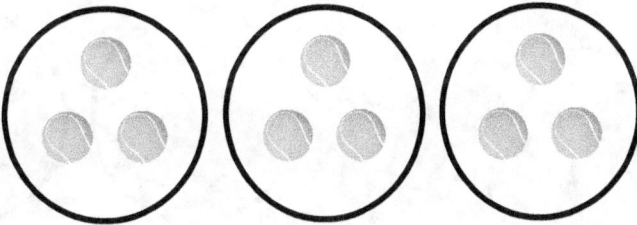

C. $\boxed{36 \div 12 = 3}$

www.math-knots.com

4. Lets learn 48 ÷ 12 = 4

A.

B.

C. $48 \div 12 = 4$

5. Lets learn $60 \div 12 = 5$

A.

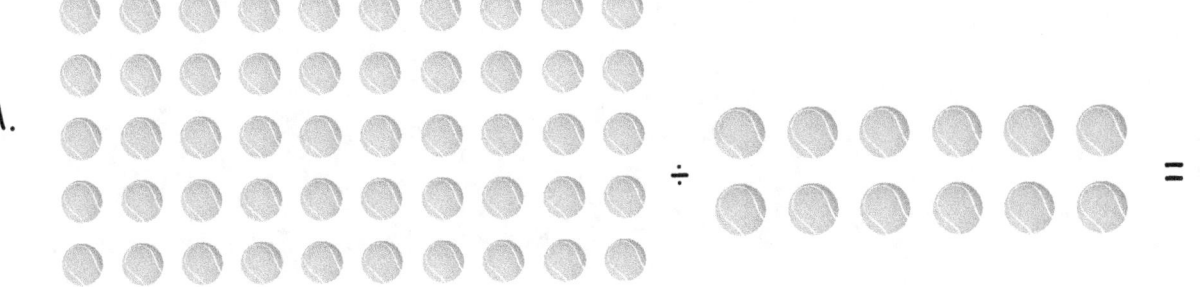

B.

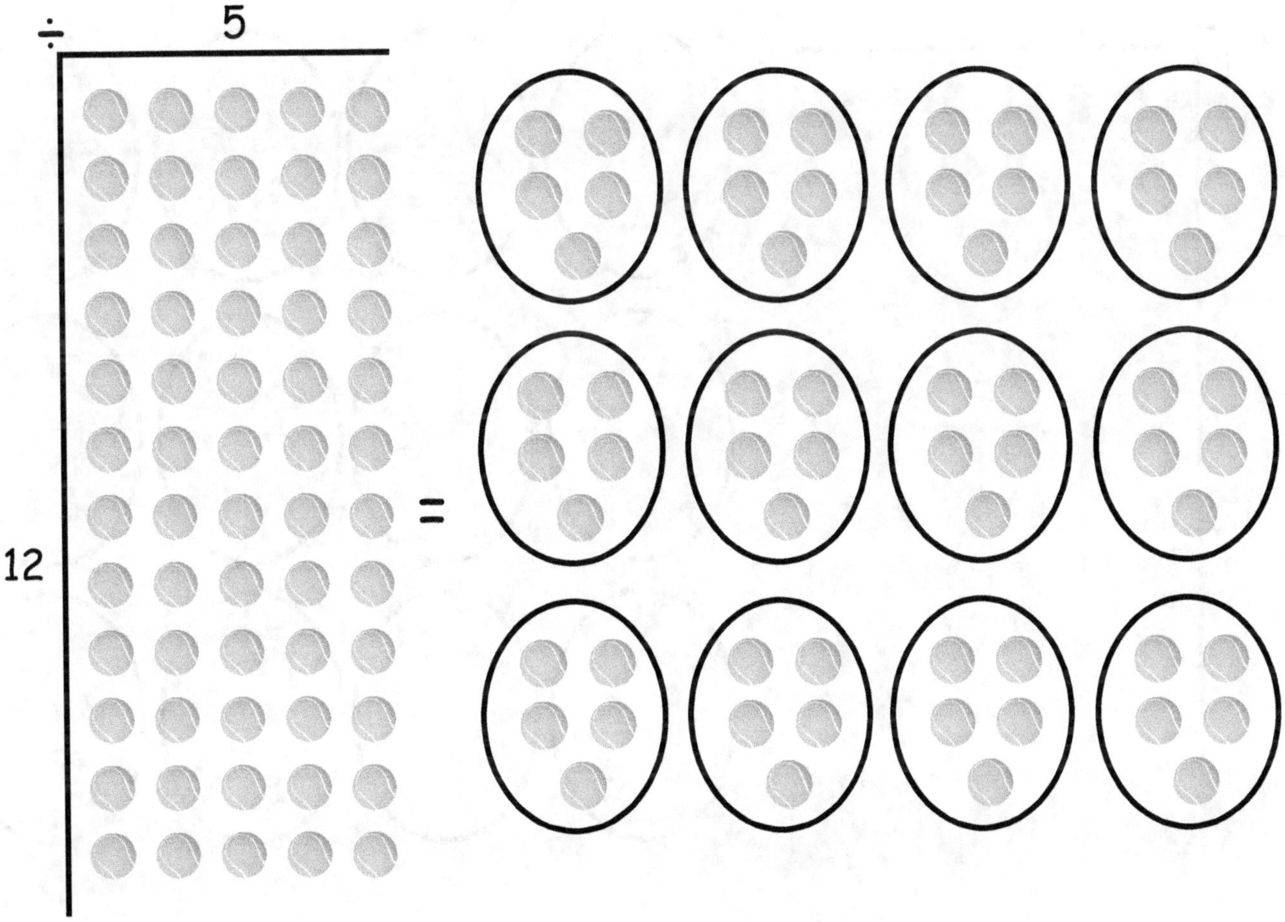

$\div$ 5

12

=

C. | $60 \div 12 = 5$ |

www.math-knots.com

6. Lets learn 72 ÷ 12 = 6

A.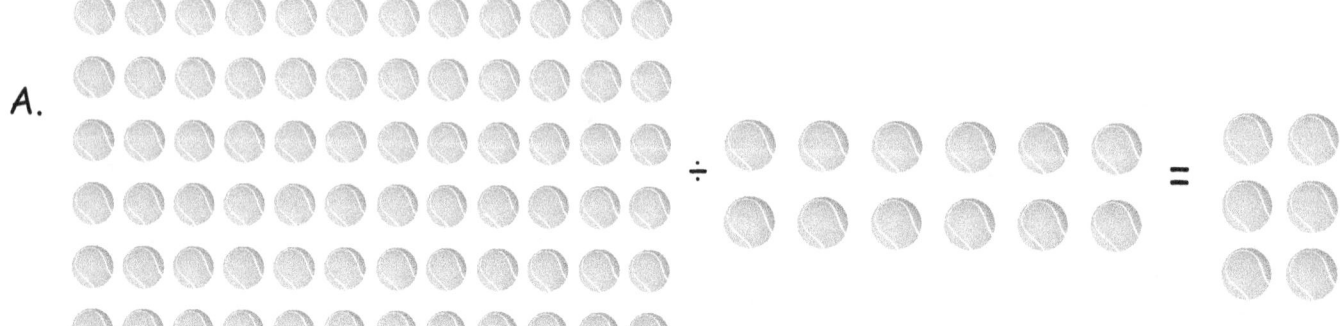

B. ÷   6

    12

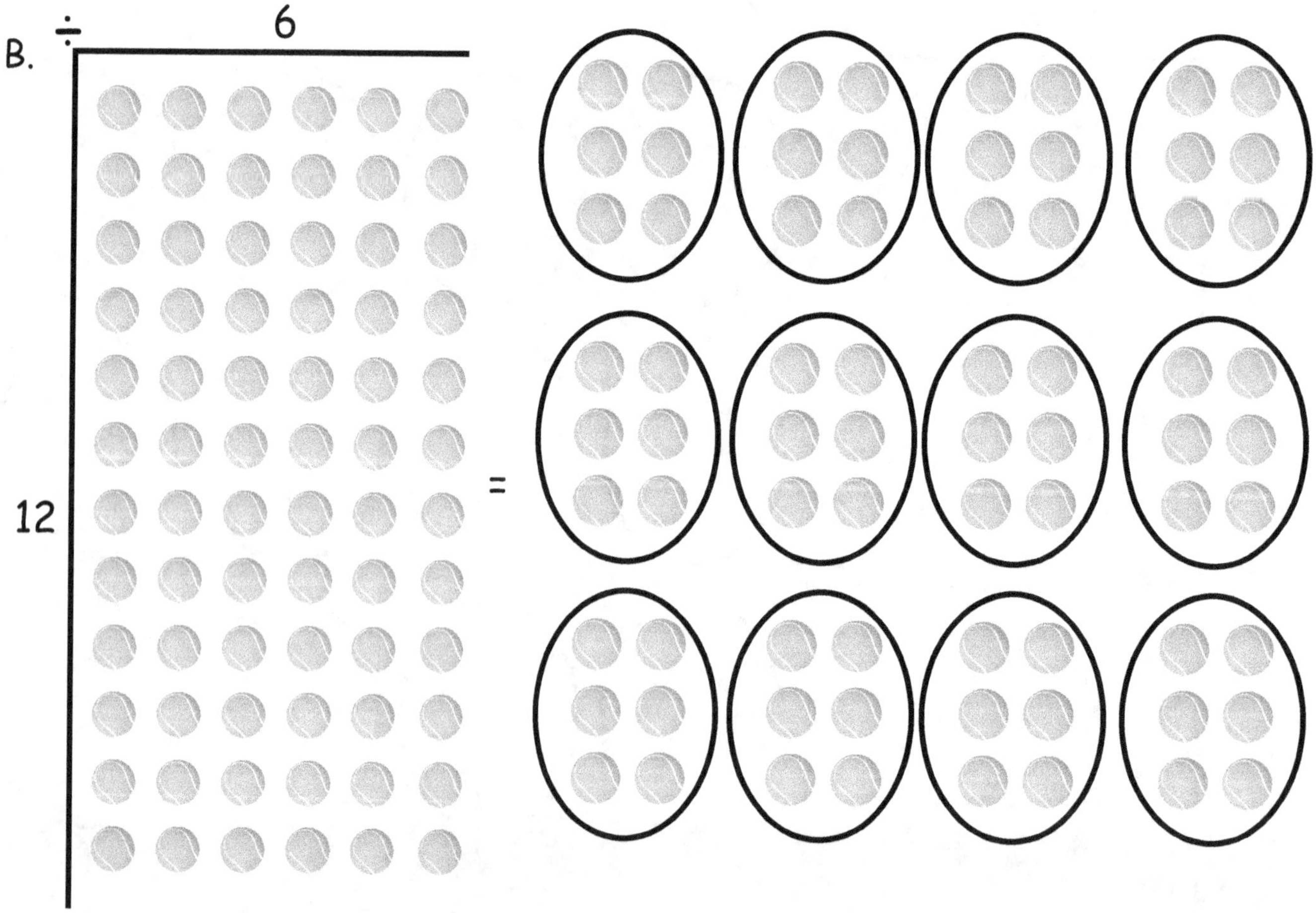

C. $72 \div 12 = 6$

www.math-knots.com

7. Lets learn $84 \div 12 = 7$

A.

B.

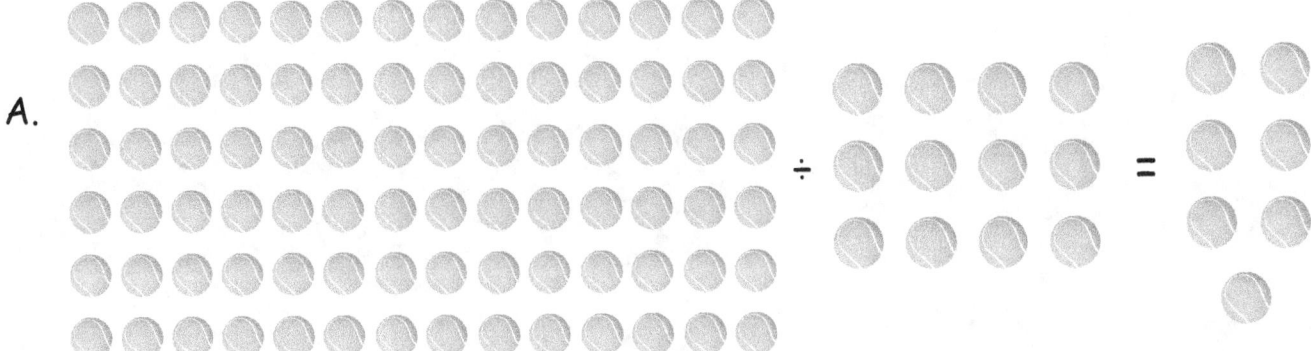

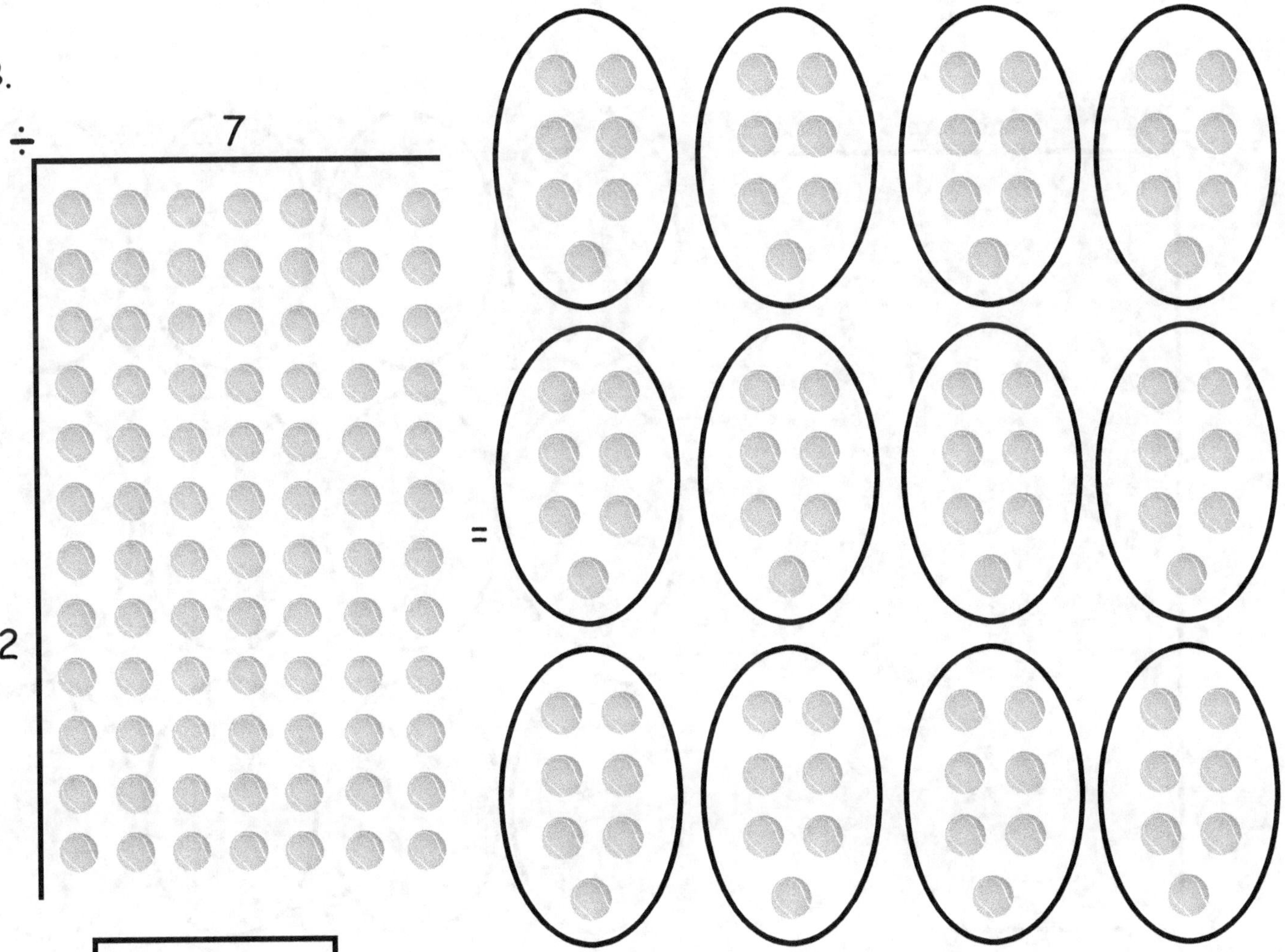

C. $84 \div 12 = 7$

8. Lets learn 96 ÷ 12 = 8

A.

B.

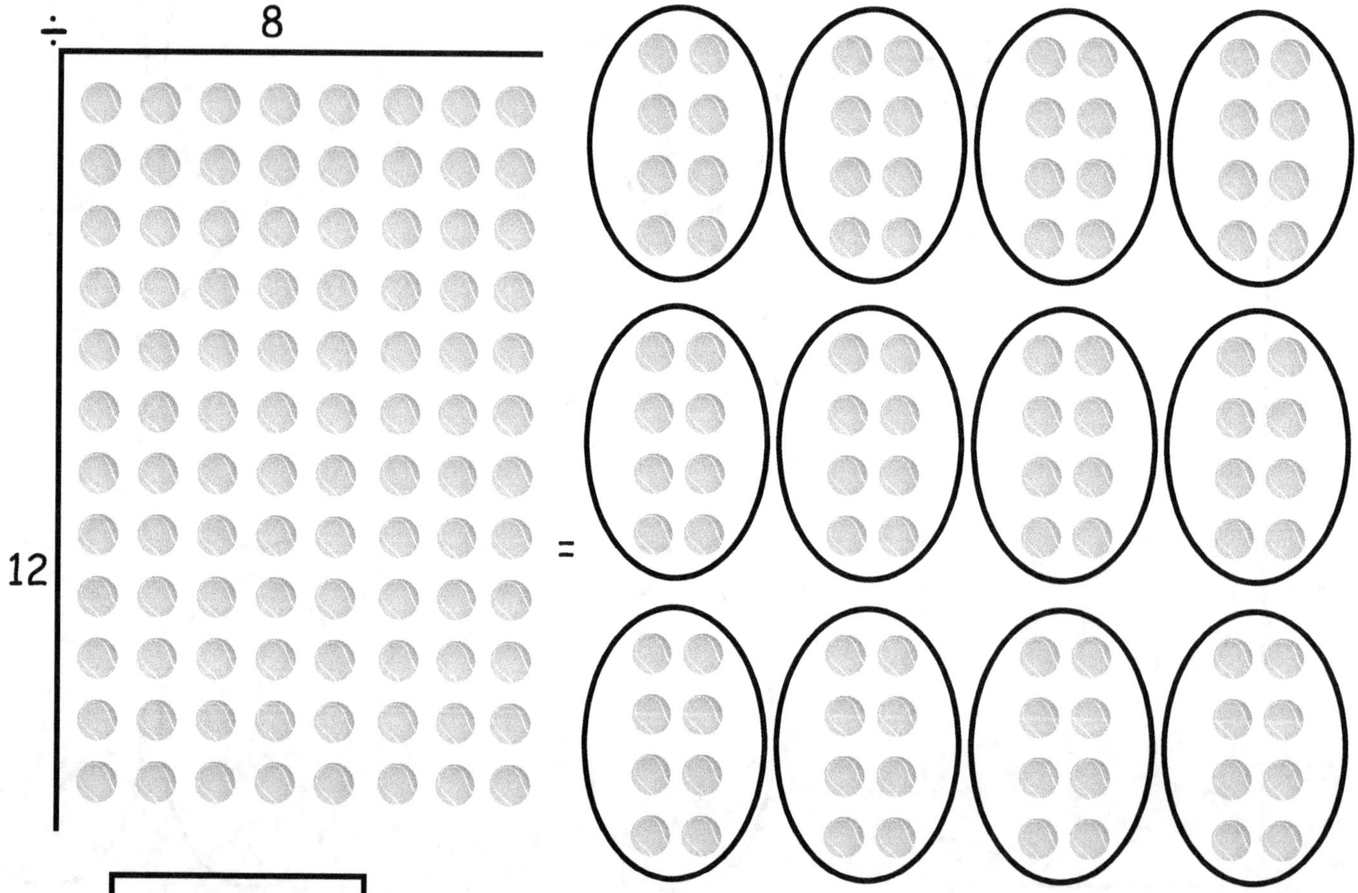

C. $96 \div 12 = 8$

9. Lets learn 108 ÷ 12 = 9

A.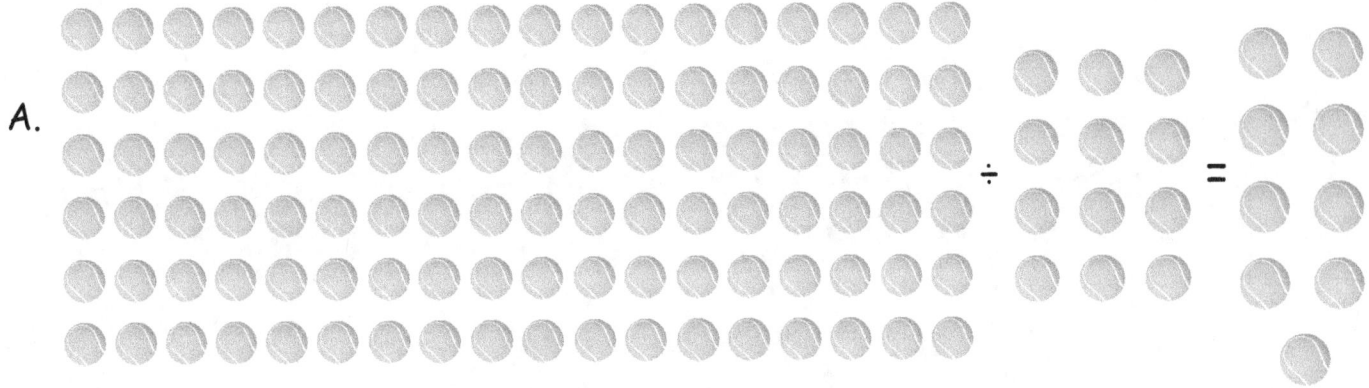

$$\div \quad = $$

B.

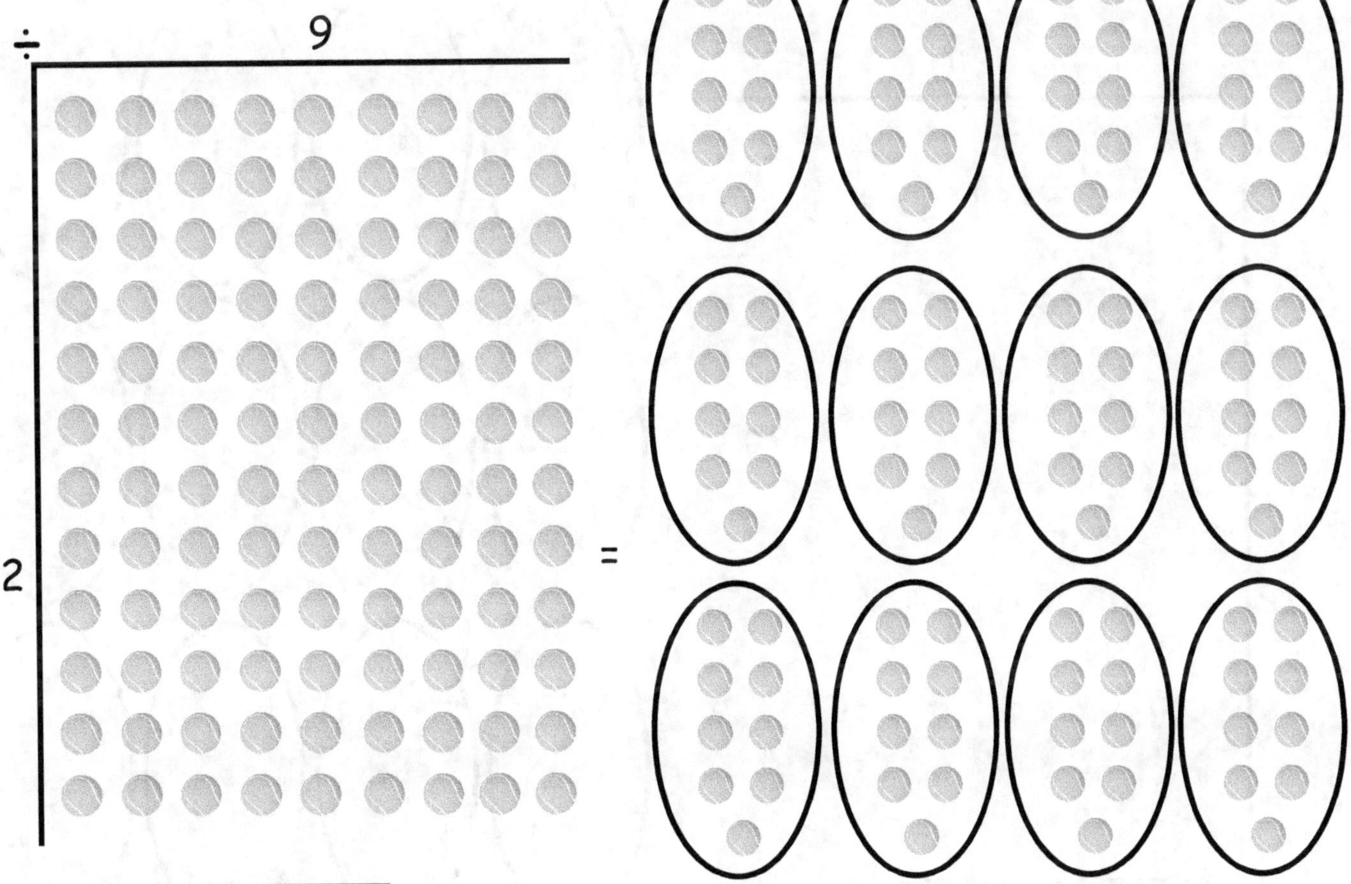

$$\div \quad 9$$

12

$$=$$

C. | 108 ÷ 12 = 9 |

10. Lets learn 120 ÷ 12 = 10

A.

B.

$\div$ 10

12

=

C. $\boxed{120 \div 12 = 10}$

11. Lets learn 132 ÷ 12 = 11

A.

B.

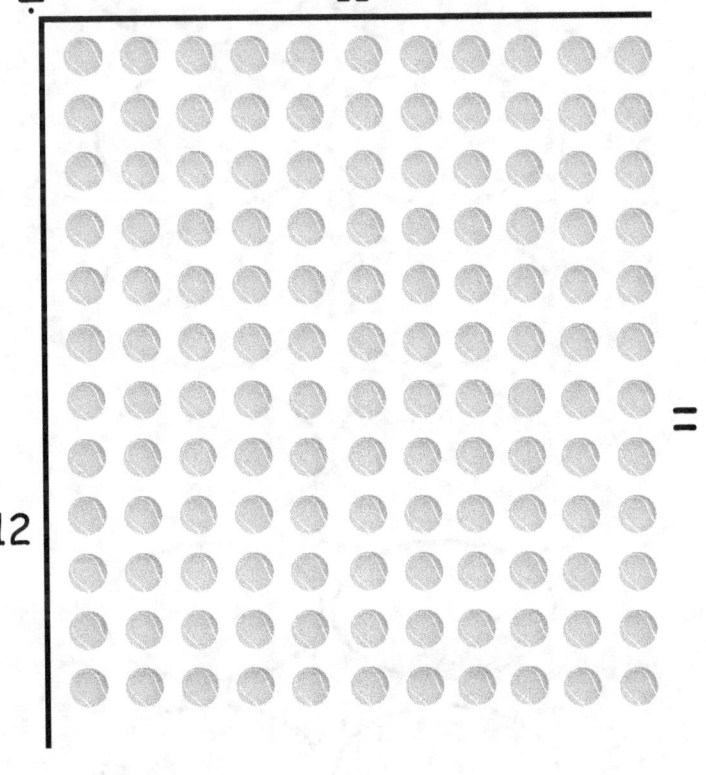

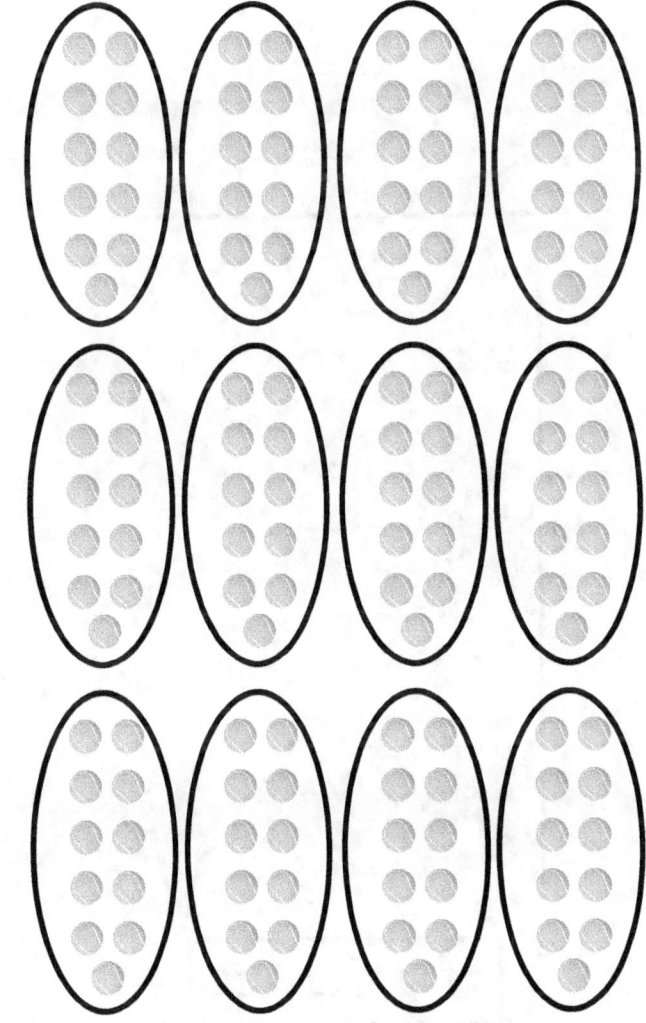

C. $132 \div 12 = 11$

www.math-knots.com

12. Lets learn 144 ÷ 12 = 12

A.

B.
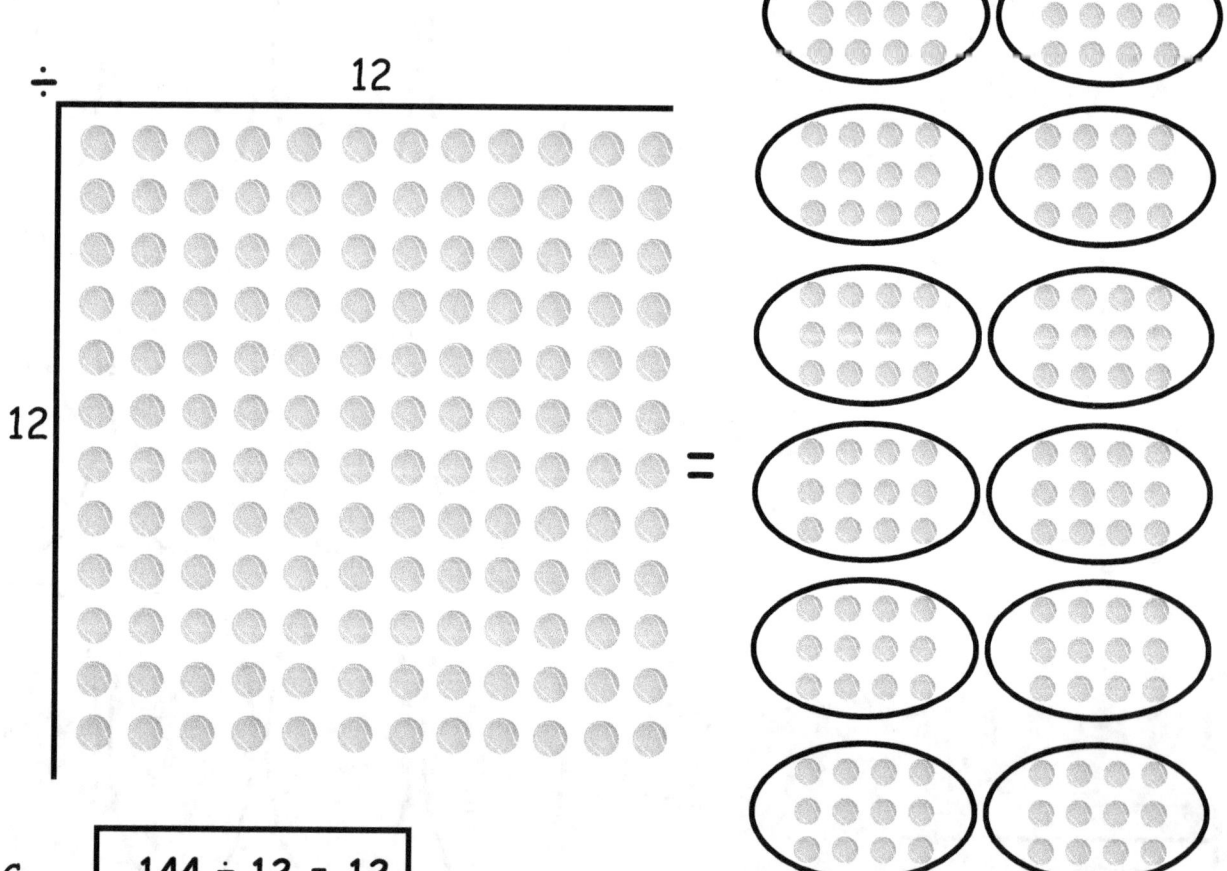

C. **144 ÷ 12 = 12**

www.math-knots.com

Exercise - 1

(A) $12\overline{)12}$ (F) $12\overline{)72}$ (K) $12\overline{)132}$

(B) $12\overline{)24}$ (G) $12\overline{)84}$ (L) $12\overline{)144}$

(C) $12\overline{)36}$ (H) $12\overline{)96}$ (M) $12\overline{)156}$

(D) $12\overline{)48}$ (I) $12\overline{)108}$ (N) $12\overline{)168}$

(E) $12\overline{)60}$ (J) $12\overline{)120}$ (O) $12\overline{)180}$

www.math-knots.com

Exercise - 2

1. 12 ÷ 12 = _____

2. 24 ÷ 12 = _____

3. 36 ÷ 12 = _____

4. 48 ÷ 12 = _____

5. 60 ÷ 12 = _____

6. 72 ÷ 12 = _____

7. 84 ÷ 12 = _____

8. 96 ÷ 12 = _____

9. 108 ÷ 12 = _____

10. 120 ÷ 12 = _____

11. 132 ÷ 12 = _____

12. 144 ÷ 12 = _____

1 × _____ = 12

2 × _____ = 24

3 × _____ = 36

4 × _____ = 48

5 × _____ = 60

6 × _____ = 72

7 × _____ = 84

8 × _____ = 96

9 × _____ = 108

10 × _____ = 120

11 × _____ = 132

12 × _____ = 144

Did You Know...?

Did you know division is splitting a number up by any give number.

www.math-knots.com

 **Exercise - 3**

1.   I am a number, I divide myself, into one equal group of 12. What am I ?

   (A)   12   (B)   1

   (C)   2   (D)   3

2.   I am a number, I divide myself, into twelve equal groups of 1. What am I ?

   (A)   12   (B)   6

   (C)   4   (D)   2

3.   I am a number, I divide myself, into twelve equal groups of 2. What am I ?

   (A)   16   (B)   12

   (C)   24   (D)   4

4.   I am a number, I divide myself, into twelve equal groups of 3. What am I ?

   (A)   3   (B)   24

   (C)   18   (D)   36

5.   I am a number, I divide myself, into twelve equal groups of 4. What am I ?

   (A)   24   (B)   4

   (C)   48   (D)   6

                                       www.math-knots.com

6. I am a number, I divide myself, into twelve equal groups of 5. What am I ?

(A) 60

(B) 12

(C) 30

(D) 5

7. I am a number, I divide myself, into twelve equal groups of 6. What am I ?

(A) 36

(B) 6

(C) 72

(D) 4

8. I am a number, I divide myself, into twelve equal groups of 7. What am I ?

(A) 12

(B) 84

(C) 7

(D) 42

9. I am a number, I divide myself, into twelve equal groups of 8. What am I ?

(A) 12

(B) 96

(C) 8

(D) 48

10. I am a number, I divide myself, into twelve equal groups of 9. What am I ?

(A) 36

(B) 9

(C) 108

(D) 54

11. I am a number, I divide myself, into twelve equal groups of 10. What am I ?

(A) 10   (B) 60

(C) 15   (D) 120

12. I am a number, I divide myself, into twelve equal groups of 11. What am I ?

(A) 66   (B) 11

(C) 132   (D) 22

13. I am a number, I divide myself, into twelve equal groups of 12. What am I ?

(A) 12   (B) 144

(C) 72   (D) 48

14. I am a number, I divide myself, into twelve equal groups of 13. What am I ?

(A) 36   (B) 13

(C) 12   (D) 156

15. I am a number, I divide myself, into twelve equal groups of 14. What am I ?

(A) 168   (B) 12

(C) 14   (D) 70

                                       www.math-knots.com

Exercise - 4

Solve the maze run below.

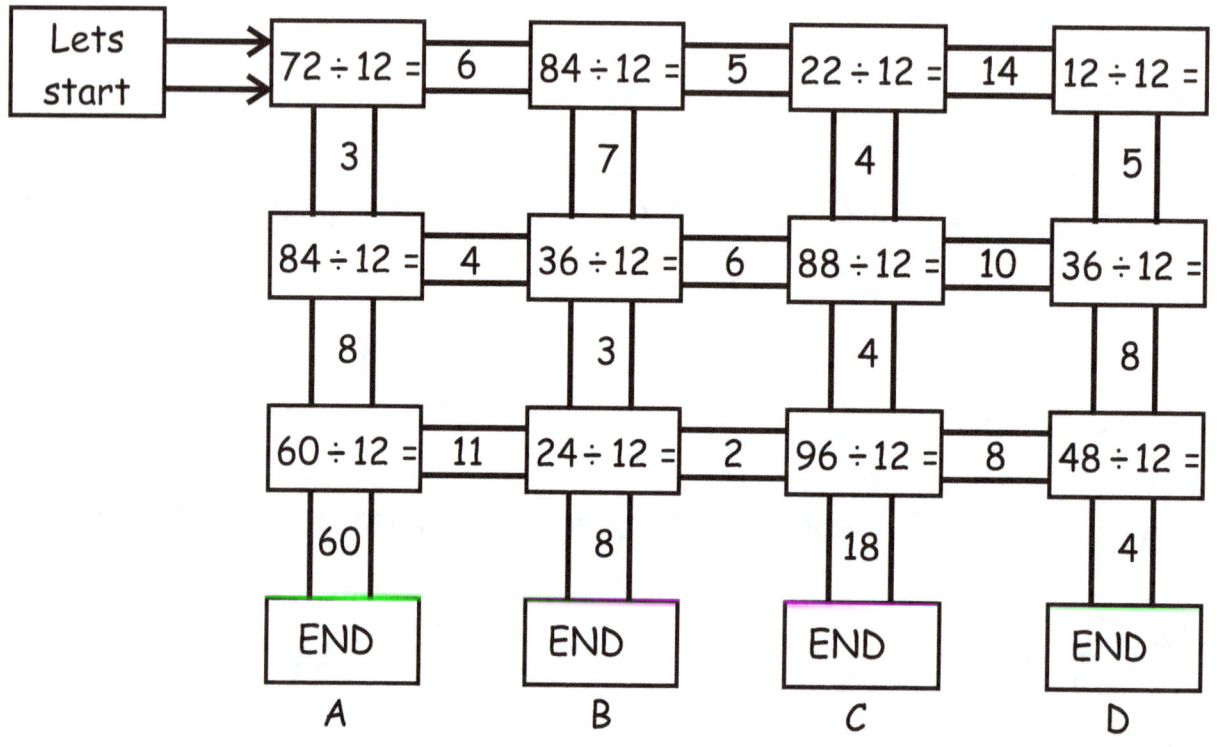

Lets start	72 ÷ 12 =	6	84 ÷ 12 =	5	22 ÷ 12 =	14	12 ÷ 12 =
	3		7		4		5
	84 ÷ 12 =	4	36 ÷ 12 =	6	88 ÷ 12 =	10	36 ÷ 12 =
	8		3		4		8
	60 ÷ 12 =	11	24 ÷ 12 =	2	96 ÷ 12 =	8	48 ÷ 12 =
	60		8		18		4
	END A		END B		END C		END D

Who won the race ? _____

www.math-knots.com

Exercise - 5

1. $12 ÷ \boxed{} = 1$ then $\boxed{} =$ _____

2. $24 ÷ \boxed{} = 12$ then $\boxed{} =$ _____

3. $36 ÷ \boxed{} = 12$ then $\boxed{} =$ _____

4. $48 ÷ \boxed{} = 12$ then $\boxed{} =$ _____

5. $60 ÷ \boxed{} = 12$ then $\boxed{} =$ _____

6. $72 ÷ \boxed{} = 12$ then $\boxed{} =$ _____

7. $84 ÷ \boxed{} = 12$ then $\boxed{} =$ _____

8. $96 ÷ \boxed{} = 12$ then $\boxed{} =$ _____

9. $108 ÷ \boxed{} = 12$ then $\boxed{} =$ _____

10. $120 ÷ \boxed{} = 12$ then $\boxed{} =$ _____

11. $132 ÷ \boxed{} = 12$ then $\boxed{} =$ _____

12. $144 ÷ \boxed{} = 12$ then $\boxed{} =$ _____

Hey you are an expert of division facts of #12 !!!

235   www.math-knots.com

www.math-knots.com

DP Exercise 1

1. 1 ÷ 1 = _____

2. 2 ÷ 1 = _____

3. 3 ÷ 1 = _____

4. 4 ÷ 1 = _____

5. 5 ÷ 1 = _____

6. 6 ÷ 1 = _____

7. 7 ÷ 1 = _____

8. 8 ÷ 1 = _____

9. 9 ÷ 1 = _____

10. 10 ÷ 1 = _____

11. 11 ÷ 1 = _____

12. 12 ÷ 1 = _____

Did you know any number when divided by
one is the number itself ?

www.math-knots.com

DP Exercise 2

1. $2 \div 2 =$ _____

2. $4 \div 2 =$ _____

3. $6 \div 2 =$ _____

4. $8 \div 2 =$ _____

5. $10 \div 2 =$ _____

6. $12 \div 2 =$ _____

7. $14 \div 2 =$ _____

8. $16 \div 2 =$ _____

9. $18 \div 2 =$ _____

10. $20 \div 2 =$ _____

11. $22 \div 2 =$ _____

12. $24 \div 2 =$ _____

1. $\times$ _____ $= 2$

2. $\times$ _____ $= 4$

3. $\times$ _____ $= 6$

4. $\times$ _____ $= 8$

5. $\times$ _____ $= 10$

6. $\times$ _____ $= 12$

7. $\times$ _____ $= 14$

8. $\times$ _____ $= 16$

9. $\times$ _____ $= 18$

10. $\times$ _____ $= 20$

11. $\times$ _____ $= 22$

12. $\times$ _____ $= 24$

Did you know division by 2 means
dividing the given number into 2 equal halfs ?

www.math-knots.com

DP Exercise 3

1.	3 ÷ 3 = _____
2.	6 ÷ 3 = _____
3.	9 ÷ 3 = _____
4.	12 ÷ 3 = _____
5.	15 ÷ 3 = _____
6.	18 ÷ 3 = _____
7.	21 ÷ 3 = _____
8.	24 ÷ 3 = _____
9.	27 ÷ 3 = _____
10.	30 ÷ 3 = _____
11.	33 ÷ 3 = _____
12.	36 ÷ 3 = _____

1	× _____ = 3
2	× _____ = 6
3	× _____ = 9
4	× _____ = 12
5	× _____ = 15
6	× _____ = 18
7	× _____ = 21
8	× _____ = 24
9	× _____ = 27
10	× _____ = 30
11	× _____ = 33
12	× _____ = 36

Did you know division by 3 means
dividing the given number into 3 equal halfs ?

www.math-knots.com

DP Exercise 4

1.	4 ÷ 4 =	_____	
2.	8 ÷ 4 =	_____	
3.	12 ÷ 4 =	_____	
4.	16 ÷ 4 =	_____	
5.	20 ÷ 4 =	_____	
6.	24 ÷ 4 =	_____	
7.	28 ÷ 4 =	_____	
8.	32 ÷ 4 =	_____	
9.	36 ÷ 4 =	_____	
10.	40 ÷ 4 =	_____	
11.	44 ÷ 4 =	_____	
12.	48 ÷ 4 =	_____	

1	× ____	=	4
2	× ____	=	8
3	× ____	=	12
4	× ____	=	16
5	× ____	=	20
6	× ____	=	24
7	× ____	=	28
8	× ____	=	32
9	× ____	=	36
10	× ____	=	40
11	× ____	=	44
12	× ____	=	48

Did you know division by 4 means
dividing the given number into 4 equal halfs ?

www.math-knots.com

DP Exercise 5

1.	5 ÷ 5	=	_____
2.	10 ÷ 5	=	_____
3.	15 ÷ 5	=	_____
4.	20 ÷ 5	=	_____
5.	25 ÷ 5	=	_____
6.	30 ÷ 5	=	_____
7.	35 ÷ 5	=	_____
8.	40 ÷ 5	=	_____
9.	45 ÷ 5	=	_____
10.	50 ÷ 5	=	_____
11.	55 ÷ 5	=	_____
12.	60 ÷ 5	=	_____

1	×	_____	=	5
2	×	_____	=	10
3	×	_____	=	15
4	×	_____	=	20
5	×	_____	=	25
6	×	_____	=	30
7	×	_____	=	35
8	×	_____	=	40
9	×	_____	=	45
10	×	_____	=	50
11	×	_____	=	55
12	×	_____	=	60

Did You Know...?

Did you know division by 5 means
dividing the given number into 5 equal halfs ?

www.math-knots.com

DP Exercise 6

1.	6 ÷ 6	=	_____
2.	12 ÷ 6	=	_____
3.	18 ÷ 6	=	_____
4.	24 ÷ 6	=	_____
5.	30 ÷ 6	=	_____
6.	36 ÷ 6	=	_____
7.	42 ÷ 6	=	_____
8.	48 ÷ 6	=	_____
9.	54 ÷ 6	=	_____
10.	60 ÷ 6	=	_____
11.	66 ÷ 6	=	_____
12.	72 ÷ 6	=	_____

1	× _____	=	6
2	× _____	=	12
3	× _____	=	18
4	× _____	=	24
5	× _____	=	30
6	× _____	=	36
7	× _____	=	42
8	× _____	=	48
9	× _____	=	54
10	× _____	=	60
11	× _____	=	66
12	× _____	=	72

Did You Know...?

Did you know division by 6 means
dividing the given number into 6 equal halfs ?

www.math-knots.com

DP Exercise 7

1.	7 ÷ 7 = _____	1 × ____ = 7
2.	14 ÷ 7 = _____	2 × ____ = 14
3.	21 ÷ 7 = _____	3 × ____ = 21
4.	28 ÷ 7 = _____	4 × ____ = 28
5.	35 ÷ 7 = _____	5 × ____ = 35
6.	42 ÷ 7 = _____	6 × ____ = 42
7.	49 ÷ 7 = _____	7 × ____ = 49
8.	56 ÷ 7 = _____	8 × ____ = 56
9.	63 ÷ 7 = _____	9 × ____ = 63
10.	70 ÷ 7 = _____	10 × ____ = 70
11.	77 ÷ 7 = _____	11 × ____ = 77
12.	84 ÷ 7 = _____	12 × ____ = 84

Did you know...?

Did you know division by 7 means
dividing the given number into 7 equal halfs ?

www.math-knots.com

DP Exercise 8

1.	8 ÷ 8 =	_____
2.	16 ÷ 8 =	_____
3.	24 ÷ 8 =	_____
4.	32 ÷ 8 =	_____
5.	40 ÷ 8 =	_____
6.	48 ÷ 8 =	_____
7.	56 ÷ 8 =	_____
8.	64 ÷ 8 =	_____
9.	72 ÷ 8 =	_____
10.	80 ÷ 8 =	_____
11.	88 ÷ 8 =	_____
12.	96 ÷ 8 =	_____

1	× ____ =	8
2	× ____ =	16
3	× ____ =	24
4	× ____ =	32
5	× ____ =	40
6	× ____ =	48
7	× ____ =	56
8	× ____ =	64
9	× ____ =	72
10	× ____ =	80
11	× ____ =	88
12	× ____ =	96

Did You Know...?

Did you know division by 8 means
dividing the given number into 8 equal halfs ?

www.math-knots.com

DP Exercise 9

1.	9 ÷ 9 =	_____
2.	18 ÷ 9 =	_____
3.	27 ÷ 9 =	_____
4.	36 ÷ 9 =	_____
5.	45 ÷ 9 =	_____
6.	54 ÷ 9 =	_____
7.	63 ÷ 9 =	_____
8.	72 ÷ 9 =	_____
9.	81 ÷ 9 =	_____
10.	90 ÷ 9 =	_____
11.	99 ÷ 9 =	_____
12.	108 ÷ 9 =	_____

1	× _____ =	9
2	× _____ =	18
3	× _____ =	27
4	× _____ =	36
5	× _____ =	45
6	× _____ =	54
7	× _____ =	63
8	× _____ =	72
9	× _____ =	81
10	× _____ =	90
11	× _____ =	99
12	× _____ =	108

Did You Know...?

Did you know division by 9 means dividing the given number into 9 equal halfs ?

www.math-knots.com

DP Exercise 10

1.	10 ÷ 10 = _____	
2.	20 ÷ 10 = _____	
3.	30 ÷ 10 = _____	
4.	40 ÷ 10 = _____	
5.	50 ÷ 10 = _____	
6.	60 ÷ 10 = _____	
7.	70 ÷ 10 = _____	
8.	80 ÷ 10 = _____	
9.	90 ÷ 10 = _____	
10.	100 ÷ 10 = _____	
11.	110 ÷ 10 = _____	
12.	120 ÷ 10 = _____	

1   × _____ = 10

2   × _____ = 20

3   × _____ = 30

4   × _____ = 40

5   × _____ = 50

6   × _____ = 60

7   × _____ = 70

8   × _____ = 80

9   × _____ = 90

10 × _____ = 100

11 × _____ = 110

12 × _____ = 120

Did you know division by 10 means
dividing the given number into 10 equal halfs ?

www.math-knots.com

DP Exercise 11

1.	$11 \div 11 =$	_____
2.	$22 \div 11 =$	_____
3.	$33 \div 11 =$	_____
4.	$44 \div 11 =$	_____
5.	$55 \div 11 =$	_____
6.	$66 \div 11 =$	_____
7.	$77 \div 11 =$	_____
8.	$88 \div 11 =$	_____
9.	$99 \div 11 =$	_____
10.	$110 \div 11 =$	_____
11.	$121 \div 11 =$	_____
12.	$132 \div 11 =$	_____

1	$\times$ _____ $= 11$
2	$\times$ _____ $= 22$
3	$\times$ _____ $= 33$
4	$\times$ _____ $= 44$
5	$\times$ _____ $= 55$
6	$\times$ _____ $= 66$
7	$\times$ _____ $= 77$
8	$\times$ _____ $= 88$
9	$\times$ _____ $= 99$
10	$\times$ _____ $= 110$
11	$\times$ _____ $= 121$
12	$\times$ _____ $= 132$

Did You Know...?

Did you know division by 11 means
dividing the given number into 11 equal halfs ?

www.math-knots.com

DP Exercise 12

1. $12 \div 12 =$ _____

2. $24 \div 12 =$ _____

3. $36 \div 12 =$ _____

4. $48 \div 12 =$ _____

5. $60 \div 12 =$ _____

6. $72 \div 12 =$ _____

7. $84 \div 12 =$ _____

8. $96 \div 12 =$ _____

9. $108 \div 12 =$ _____

10. $120 \div 12 =$ _____

11. $132 \div 12 =$ _____

12. $144 \div 12 =$ _____

$1 \times$ _____ $= 12$

$2 \times$ _____ $= 24$

$3 \times$ _____ $= 36$

$4 \times$ _____ $= 48$

$5 \times$ _____ $= 60$

$6 \times$ _____ $= 72$

$7 \times$ _____ $= 84$

$8 \times$ _____ $= 96$

$9 \times$ _____ $= 108$

$10 \times$ _____ $= 120$

$11 \times$ _____ $= 132$

$12 \times$ _____ $= 144$

Did You Know...?

Did you know division by 12 means
dividing the given number into 12 equal halfs ?

www.math-knots.com

DP Exercise 13

1.	$13 \div 13 =$	_____
2.	$26 \div 13 =$	_____
3.	$39 \div 13 =$	_____
4.	$52 \div 13 =$	_____
5.	$65 \div 13 =$	_____
6.	$78 \div 13 =$	_____
7.	$91 \div 13 =$	_____
8.	$104 \div 13 =$	_____
9.	$117 \div 13 =$	_____
10.	$130 \div 13 =$	_____
11.	$143 \div 13 =$	_____
12.	$156 \div 13 =$	_____

1	$\times$ _____	$= 13$
2	$\times$ _____	$= 26$
3	$\times$ _____	$= 39$
4	$\times$ _____	$= 52$
5	$\times$ _____	$= 65$
6	$\times$ _____	$= 78$
7	$\times$ _____	$= 91$
8	$\times$ _____	$= 104$
9	$\times$ _____	$= 117$
10	$\times$ _____	$= 130$
11	$\times$ _____	$= 143$
12	$\times$ _____	$= 156$

Did you know division by 13 means
dividing the given number into 13 equal halfs ?

www.math-knots.com

DP Exercise 14

1.	14 ÷ 14 = _____
2.	28 ÷ 14 = _____
3.	42 ÷ 14 = _____
4.	56 ÷ 14 = _____
5.	70 ÷ 14 = _____
6.	84 ÷ 14 = _____
7.	98 ÷ 14 = _____
8.	112 ÷ 14 = _____
9.	126 ÷ 14 = _____
10.	140 ÷ 14 = _____
11.	154 ÷ 14 = _____
12.	168 ÷ 14 = _____

1	× _____ = 14
2	× _____ = 28
3	× _____ = 42
4	× _____ = 56
5	× _____ = 70
6	× _____ = 84
7	× _____ = 98
8	× _____ = 112
9	× _____ = 126
10	× _____ = 140
11	× _____ = 154
12	× _____ = 168

Did you know division by 14 means
dividing the given number into 14 equal halfs ?

www.math-knots.com

DP Exercise 15

1.	$15 \div 15 =$ _____
2.	$30 \div 15 =$ _____
3.	$45 \div 15 =$ _____
4.	$60 \div 15 =$ _____
5.	$75 \div 15 =$ _____
6.	$90 \div 15 =$ _____
7.	$105 \div 15 =$ _____
8.	$120 \div 15 =$ _____
9.	$135 \div 15 =$ _____
10.	$150 \div 15 =$ _____
11.	$165 \div 15 =$ _____
12.	$180 \div 15 =$ _____

1	$\times$ _____ $= 15$
2	$\times$ _____ $= 30$
3	$\times$ _____ $= 45$
4	$\times$ _____ $= 60$
5	$\times$ _____ $= 75$
6	$\times$ _____ $= 90$
7	$\times$ _____ $= 105$
8	$\times$ _____ $= 120$
9	$\times$ _____ $= 135$
10	$\times$ _____ $= 150$
11	$\times$ _____ $= 165$
12	$\times$ _____ $= 180$

Did you know division by 15 means
dividing the given number into 15 equal halfs ?

www.math-knots.com

DP Exercise 16

1.	16 ÷ 16 = _____
2.	32 ÷ 16 = _____
3.	48 ÷ 16 = _____
4.	64 ÷ 16 = _____
5.	80 ÷ 16 = _____
6.	96 ÷ 16 = _____
7.	112 ÷ 16 = _____
8.	128 ÷ 16 = _____
9.	144 ÷ 16 = _____
10.	160 ÷ 16 = _____
11.	176 ÷ 16 = _____
12.	192 ÷ 16 = _____

1	× _____ = 16
2	× _____ = 32
3	× _____ = 48
4	× _____ = 64
5	× _____ = 80
6	× _____ = 96
7	× _____ = 112
8	× _____ = 128
9	× _____ = 144
10	× _____ = 160
11	× _____ = 176
12	× _____ = 192

Did you know division by 16 means
dividing the given number into 16 equal halfs ?

www.math-knots.com

DP Exercise 17

1.	17 ÷ 17 = _____
2.	34 ÷ 17 = _____
3.	51 ÷ 17 = _____
4.	68 ÷ 17 = _____
5.	85 ÷ 17 = _____
6.	102 ÷ 17 = _____
7.	119 ÷ 17 = _____
8.	136 ÷ 17 = _____
9.	153 ÷ 17 = _____
10.	170 ÷ 17 = _____
11.	187 ÷ 17 = _____
12.	204 ÷ 17 = _____

1	× _____ = 17
2	× _____ = 34
3	× _____ = 51
4	× _____ = 68
5	× _____ = 85
6	× _____ = 102
7	× _____ = 119
8	× _____ = 136
9	× _____ = 153
10	× _____ = 170
11	× _____ = 187
12	× _____ = 204

Did you know division by 17 means
dividing the given number into 17 equal halfs ?

www.math-knots.com

DP Exercise 18

1.	$18 \div 18 =$ _____
2.	$36 \div 18 =$ _____
3.	$54 \div 18 =$ _____
4.	$72 \div 18 =$ _____
5.	$90 \div 18 =$ _____
6.	$108 \div 18 =$ _____
7.	$126 \div 18 =$ _____
8.	$144 \div 18 =$ _____
9.	$162 \div 18 =$ _____
10.	$180 \div 18 =$ _____
11.	$198 \div 18 =$ _____
12.	$216 \div 18 =$ _____

1	$\times$ ____ $= 18$
2	$\times$ ____ $= 36$
3	$\times$ ____ $= 54$
4	$\times$ ____ $= 72$
5	$\times$ ____ $= 90$
6	$\times$ ____ $= 108$
7	$\times$ ____ $= 126$
8	$\times$ ____ $= 142$
9	$\times$ ____ $= 162$
10	$\times$ ____ $= 180$
11	$\times$ ____ $= 198$
12	$\times$ ____ $= 216$

Did You Know...?

Did you know division by 18 means dividing the given number into 18 equal halfs ?

www.math-knots.com

DP Exercise 19

1.	19 ÷ 19 =	_____
2.	38 ÷ 19 =	_____
3.	57 ÷ 19 =	_____
4.	76 ÷ 19 =	_____
5.	95 ÷ 19 =	_____
6.	114 ÷ 19 =	_____
7.	133 ÷ 19 =	_____
8.	152 ÷ 19 =	_____
9.	171 ÷ 19 =	_____
10.	190 ÷ 19 =	_____
11.	209 ÷ 19 =	_____
12.	228 ÷ 19 =	_____

1 × _____ = 19
2 × _____ = 38
3 × _____ = 57
4 × _____ = 76
5 × _____ = 95
6 × _____ = 114
7 × _____ = 133
8 × _____ = 152
9 × _____ = 171
10 × _____ = 190
11 × _____ = 209
12 × _____ = 228

Did you know division by 19 means dividing the given number into 19 equal halfs ?

www.math-knots.com

DP Exercise 20

1.	20 ÷ 20	=	_____
2.	40 ÷ 20	=	_____
3.	60 ÷ 20	=	_____
4.	80 ÷ 20	=	_____
5.	100 ÷ 20	=	_____
6.	120 ÷ 20	=	_____
7.	140 ÷ 20	=	_____
8.	160 ÷ 20	=	_____
9.	180 ÷ 20	=	_____
10.	200 ÷ 20	=	_____
11.	220 ÷ 20	=	_____
12.	240 ÷ 20	=	_____

1	× ____	=	20
2	× ____	=	40
3	× ____	=	60
4	× ____	=	80
5	× ____	=	100
6	× ____	=	120
7	× ____	=	140
8	× ____	=	160
9	× ____	=	180
10	× ____	=	200
11	× ____	=	220
12	× ____	=	240

Did you know division by 20 means
dividing the given number into 20 equal halfs ?

DP Exercise 21

1. $21 \div 21 =$ _____

2. $42 \div 21 =$ _____

3. $63 \div 21 =$ _____

4. $84 \div 21 =$ _____

5. $105 \div 21 =$ _____

6. $126 \div 21 =$ _____

7. $147 \div 21 =$ _____

8. $168 \div 21 =$ _____

9. $189 \div 21 =$ _____

10. $210 \div 21 =$ _____

11. $231 \div 21 =$ _____

12. $252 \div 21 =$ _____

1 $\times$ _____ $= 21$

2 $\times$ _____ $= 42$

3 $\times$ _____ $= 63$

4 $\times$ _____ $= 84$

5 $\times$ _____ $= 105$

6 $\times$ _____ $= 126$

7 $\times$ _____ $= 147$

8 $\times$ _____ $= 168$

9 $\times$ _____ $= 189$

10 $\times$ _____ $= 210$

11 $\times$ _____ $= 231$

12 $\times$ _____ $= 252$

Did you know division by 21 means
dividing the given number into 21 equal halfs ?

www.math-knots.com

DP Exercise 22

1. 22 ÷ 22 = _____

2. 44 ÷ 22 = _____

3. 66 ÷ 22 = _____

4. 88 ÷ 22 = _____

5. 110 ÷ 22 = _____

6. 132 ÷ 22 = _____

7. 154 ÷ 22 = _____

8. 176 ÷ 22 = _____

9. 198 ÷ 22 = _____

10. 220 ÷ 22 = _____

11. 242 ÷ 22 = _____

12. 264 ÷ 22 = _____

1 × _____ = 22

2 × _____ = 44

3 × _____ = 66

4 × _____ = 88

5 × _____ = 110

6 × _____ = 132

7 × _____ = 154

8 × _____ = 176

9 × _____ = 198

10 × _____ = 220

11 × _____ = 242

12 × _____ = 264

Did you know division by 22 means dividing the given number into 22 equal halfs ?

www.math-knots.com

DP Exercise 23

1.	23 ÷ 23 =	_____
2.	46 ÷ 23 =	_____
3.	69 ÷ 23 =	_____
4.	92 ÷ 22 =	_____
5.	115 ÷ 23 =	_____
6.	138 ÷ 23 =	_____
7.	161 ÷ 23 =	_____
8.	184 ÷ 23 =	_____
9.	207 ÷ 23 =	_____
10.	230 ÷ 23 =	_____
11.	253 ÷ 23 =	_____
12.	276 ÷ 23 =	_____

1	× ____	= 23
2	× ____	= 46
3	× ____	= 69
4	× ____	= 92
5	× ____	= 115
6	× ____	= 138
7	× ____	= 161
8	× ____	= 184
9	× ____	= 207
10	× ____	= 230
11	× ____	= 253
12	× ____	= 276

Did you know division by 23 means
dividing the given number into 23 equal halfs ?

www.math-knots.com

DP Exercise 24

1. $24 \div 24 =$ _____	1 $\times$ ____ $= 24$
2. $48 \div 24 =$ _____	2 $\times$ ____ $= 48$
3. $72 \div 24 =$ _____	3 $\times$ ____ $= 72$
4. $96 \div 24 =$ _____	4 $\times$ ____ $= 96$
5. $120 \div 24 =$ _____	5 $\times$ ____ $= 120$
6. $144 \div 24 =$ _____	6 $\times$ ____ $= 144$
7. $168 \div 24 =$ _____	7 $\times$ ____ $= 168$
8. $192 \div 24 =$ _____	8 $\times$ ____ $= 192$
9. $216 \div 24 =$ _____	9 $\times$ ____ $= 216$
10. $240 \div 24 =$ _____	10 $\times$ ____ $= 240$
11. $264 \div 24 =$ _____	11 $\times$ ____ $= 264$
12. $288 \div 24 =$ _____	12 $\times$ ____ $= 288$

Did you know division by 24 means
dividing the given number into 24 equal halfs ?

www.math-knots.com

DP Exercise 25

1.	$25 \div 25 =$	_____
2.	$50 \div 25 =$	_____
3.	$75 \div 25 =$	_____
4.	$100 \div 25 =$	_____
5.	$125 \div 25 =$	_____
6.	$150 \div 25 =$	_____
7.	$175 \div 25 =$	_____
8.	$200 \div 25 =$	_____
9.	$225 \div 25 =$	_____
10.	$250 \div 25 =$	_____
11.	$275 \div 25 =$	_____
12.	$300 \div 25 =$	_____

1	$\times$ _____	$= 25$
2	$\times$ _____	$= 50$
3	$\times$ _____	$= 75$
4	$\times$ _____	$= 100$
5	$\times$ _____	$= 125$
6	$\times$ _____	$= 150$
7	$\times$ _____	$= 175$
8	$\times$ _____	$= 200$
9	$\times$ _____	$= 225$
10	$\times$ _____	$= 250$
11	$\times$ _____	$= 275$
12	$\times$ _____	$= 300$

Did you know division by 25 means
dividing the given number into 25 equal halfs ?

www.math-knots.com

DP Exercise 26

1.	67	÷ 1	=	_____
2.	44	÷ 2	=	_____
3.	99	÷ 3	=	_____
4.	80	÷ 4	=	_____
5.	95	÷ 5	=	_____
6.	126	÷ 6	=	_____
7.	49	÷ 7	=	_____
8.	96	÷ 8	=	_____
9.	189	÷ 9	=	_____
10.	800	÷ 10	=	_____
11.	1331	÷ 11	=	_____
12.	1728	÷ 12	=	_____

1	× ____	=	67
2	× ____	=	44
3	× ____	=	99
4	× ____	=	80
5	× ____	=	95
6	× ____	=	126
7	× ____	=	49
8	× ____	=	96
9	× ____	=	189
10	× ____	=	800
11	× ____	=	1331
12	× ____	=	1728

Did you know operations , division and multiplication are opposites of each other ?

www.math-knots.com

DP Exercise 27

1. 591 ÷ 1 = _____

2. 404 ÷ 2 = _____

3. 303 ÷ 3 = _____

4. 84 ÷ 4 = _____

5. 125 ÷ 5 = _____

6. 246 ÷ 6 = _____

7. 777 ÷ 7 = _____

8. 160 ÷ 8 = _____

9. 279 ÷ 9 = _____

10. 2010 ÷ 10 = _____

11. 11011 ÷ 11 = _____

12. 121212 ÷ 12 = _____

1 × _____ = 591

2 × _____ = 404

3 × _____ = 303

4 × _____ = 84

5 × _____ = 125

6 × _____ = 246

7 × _____ = 777

8 × _____ = 160

9 × _____ = 279

10 × _____ = 2010

11 × _____ = 11011

12 × _____ = 121212

Did you know operations , division and multiplication
are opposites of each other ?

www.math-knots.com

DP Exercise 28

(A) $1\overline{)2}$ (F) $1\overline{)7}$ (K) $1\overline{)12}$

(B) $1\overline{)3}$ (G) $1\overline{)8}$ (L) $1\overline{)13}$

(C) $1\overline{)4}$ (H) $1\overline{)9}$ (M) $1\overline{)14}$

(D) $1\overline{)5}$ (I) $1\overline{)10}$ (N) $1\overline{)15}$

(E) $1\overline{)6}$ (J) $1\overline{)11}$ (O) $1\overline{)16}$

www.math-knots.com

DP Exercise 29

(A) $2\overline{)2}$

(F) $2\overline{)12}$

(K) $2\overline{)22}$

(B) $2\overline{)4}$

(G) $2\overline{)14}$

(L) $2\overline{)24}$

(C) $2\overline{)6}$

(H) $2\overline{)16}$

(M) $2\overline{)26}$

(D) $2\overline{)8}$

(I) $2\overline{)18}$

(N) $2\overline{)28}$

(E) $2\overline{)10}$

(J) $2\overline{)20}$

(O) $2\overline{)30}$

www.math-knots.com

DP Exercise 30

(A) $3\overline{)3}$ (F) $3\overline{)18}$ (K) $3\overline{)33}$

(B) $3\overline{)6}$ (G) $3\overline{)21}$ (L) $3\overline{)36}$

(C) $3\overline{)9}$ (H) $3\overline{)24}$ (M) $3\overline{)39}$

(D) $3\overline{)12}$ (I) $3\overline{)27}$ (N) $3\overline{)42}$

(E) $3\overline{)15}$ (J) $3\overline{)30}$ (O) $3\overline{)45}$

DP Exercise 31

(A) 4$\overline{)4}$

(F) 4$\overline{)24}$

(K) 4$\overline{)44}$

(B) 4$\overline{)8}$

(G) 4$\overline{)28}$

(L) 4$\overline{)48}$

(C) 4$\overline{)12}$

(H) 4$\overline{)32}$

(M) 4$\overline{)52}$

(D) 4$\overline{)16}$

(I) 4$\overline{)36}$

(N) 4$\overline{)56}$

(E) 4$\overline{)20}$

(J) 4$\overline{)40}$

(O) 4$\overline{)60}$

www.math-knots.com

DP Exercise 32

(A) 5$\overline{\smash{)}5}$

(B) 5$\overline{\smash{)}10}$

(C) 5$\overline{\smash{)}15}$

(D) 5$\overline{\smash{)}20}$

(E) 5$\overline{\smash{)}25}$

(F) 5$\overline{\smash{)}30}$

(G) 5$\overline{\smash{)}35}$

(H) 5$\overline{\smash{)}40}$

(I) 5$\overline{\smash{)}45}$

(J) 5$\overline{\smash{)}50}$

(K) 5$\overline{\smash{)}55}$

(L) 5$\overline{\smash{)}60}$

(M) 5$\overline{\smash{)}65}$

(N) 5$\overline{\smash{)}70}$

(O) 5$\overline{\smash{)}75}$

www.math-knots.com

DP Exercise 33

(A) $6\overline{)6}$

(F) $6\overline{)36}$

(K) $6\overline{)66}$

(B) $6\overline{)12}$

(G) $6\overline{)42}$

(L) $6\overline{)72}$

(C) $6\overline{)18}$

(H) $6\overline{)48}$

(M) $6\overline{)78}$

(D) $6\overline{)24}$

(I) $6\overline{)54}$

(N) $6\overline{)84}$

(E) $6\overline{)30}$

(J) $6\overline{)60}$

(O) $6\overline{)90}$

www.math-knots.com

DP Exercise 34

(A) $7\overline{)7}$

(F) $7\overline{)42}$

(K) $7\overline{)77}$

(B) $7\overline{)14}$

(G) $7\overline{)49}$

(L) $7\overline{)84}$

(C) $7\overline{)21}$

(H) $7\overline{)56}$

(M) $7\overline{)91}$

(D) $7\overline{)28}$

(I) $7\overline{)63}$

(N) $7\overline{)98}$

(E) $7\overline{)35}$

(J) $7\overline{)70}$

(O) $7\overline{)105}$

DP Exercise 35

(A) 8)8   (F) 8)48   (K) 8)88

(B) 8)16   (G) 8)56   (L) 8)96

(C) 8)24   (H) 8)64   (M) 8)104

(D) 8)32   (I) 8)72   (N) 8)112

(E) 8)40   (J) 8)80   (O) 8)120

                   www.math-knots.com

DP Exercise 36

(A) $9\overline{)9}$ (F) $9\overline{)54}$ (K) $9\overline{)99}$

(B) $9\overline{)18}$ (G) $9\overline{)63}$ (L) $9\overline{)108}$

(C) $9\overline{)27}$ (H) $9\overline{)72}$ (M) $9\overline{)117}$

(D) $9\overline{)36}$ (I) $9\overline{)81}$ (N) $9\overline{)126}$

(E) $9\overline{)45}$ (J) $9\overline{)90}$ (O) $9\overline{)135}$

www.math-knots.com

DP Exercise 37

(A) $10\overline{)10}$ (F) $10\overline{)60}$ (K) $10\overline{)110}$

(B) $10\overline{)20}$ (G) $10\overline{)70}$ (L) $10\overline{)120}$

(C) $10\overline{)30}$ (H) $10\overline{)80}$ (M) $10\overline{)130}$

(D) $10\overline{)40}$ (I) $10\overline{)90}$ (N) $10\overline{)140}$

(E) $10\overline{)50}$ (J) $10\overline{)100}$ (O) $10\overline{)150}$

DP Exercise 38

(A) $11\overline{)11}$

(F) $11\overline{)66}$

(K) $11\overline{)121}$

(B) $11\overline{)22}$

(G) $11\overline{)77}$

(L) $11\overline{)132}$

(C) $11\overline{)33}$

(H) $11\overline{)88}$

(M) $11\overline{)143}$

(D) $11\overline{)44}$

(I) $11\overline{)99}$

(N) $11\overline{)154}$

(E) $11\overline{)55}$

(J) $11\overline{)110}$

(O) $11\overline{)165}$

www.math-knots.com

DP Exercise 39

(A) $12\overline{)12}$ (F) $12\overline{)72}$ (K) $12\overline{)132}$

(B) $12\overline{)24}$ (G) $12\overline{)84}$ (L) $12\overline{)144}$

(C) $12\overline{)36}$ (H) $12\overline{)96}$ (M) $12\overline{)156}$

(D) $12\overline{)48}$ (I) $12\overline{)108}$ (N) $12\overline{)168}$

(E) $12\overline{)60}$ (J) $12\overline{)120}$ (O) $12\overline{)180}$

www.math-knots.com

DP Exercise 40

(A) $13\overline{)13}$

(F) $13\overline{)78}$

(K) $13\overline{)143}$

(B) $13\overline{)26}$

(G) $13\overline{)91}$

(L) $13\overline{)156}$

(C) $13\overline{)39}$

(H) $13\overline{)104}$

(M) $13\overline{)169}$

(D) $13\overline{)52}$

(I) $13\overline{)117}$

(N) $13\overline{)182}$

(E) $13\overline{)65}$

(J) $13\overline{)130}$

(O) $13\overline{)195}$

www.math-knots.com

DP Exercise 41

(A) $14\overline{)14}$ (F) $14\overline{)84}$ (K) $14\overline{)154}$

(B) $14\overline{)28}$ (G) $14\overline{)98}$ (L) $14\overline{)168}$

(C) $14\overline{)42}$ (H) $14\overline{)112}$ (M) $14\overline{)182}$

(D) $14\overline{)56}$ (I) $14\overline{)126}$ (N) $14\overline{)196}$

(E) $14\overline{)70}$ (J) $14\overline{)140}$ (O) $14\overline{)210}$

www.math-knots.com

DP Exercise 42

(A) 15)‾15‾

(F) 15)‾90‾

(K) 15)‾165‾

(B) 15)‾30‾

(G) 15)‾105‾

(L) 15)‾180‾

(C) 15)‾45‾

(H) 15)‾120‾

(M) 15)‾195‾

(D) 15)‾60‾

(I) 15)‾135‾

(N) 15)‾210‾

(E) 15)‾75‾

(J) 15)‾150‾

(O) 15)‾225‾

www.math-knots.com

DP Exercise 43

(A) $16\overline{)16}$ (F) $16\overline{)96}$ (K) $16\overline{)176}$

(B) $16\overline{)32}$ (G) $16\overline{)112}$ (L) $16\overline{)192}$

(C) $16\overline{)48}$ (H) $16\overline{)128}$ (M) $16\overline{)208}$

(D) $16\overline{)64}$ (I) $16\overline{)144}$ (N) $16\overline{)224}$

(E) $16\overline{)80}$ (J) $16\overline{)160}$ (O) $16\overline{)240}$

www.math-knots.com

DP Exercise 44

(A) $17\overline{)17}$

(F) $17\overline{)102}$

(K) $17\overline{)187}$

(B) $17\overline{)34}$

(G) $17\overline{)119}$

(L) $17\overline{)204}$

(C) $17\overline{)51}$

(H) $17\overline{)136}$

(M) $17\overline{)221}$

(D) $17\overline{)68}$

(I) $17\overline{)153}$

(N) $17\overline{)238}$

(E) $17\overline{)85}$

(J) $17\overline{)170}$

(O) $17\overline{)255}$

www.math-knots.com

DP Exercise 45

(A) 18)‾18‾   (F) 18)‾108‾   (K) 18)‾198‾

(B) 18)‾36‾   (G) 18)‾126‾   (L) 18)‾216‾

(C) 18)‾54‾   (H) 18)‾144‾   (M) 18)‾234‾

(D) 18)‾72‾   (I) 18)‾162‾   (N) 18)‾252‾

(E) 18)‾90‾   (J) 18)‾180‾   (O) 18)‾270‾

www.math-knots.com

DP Exercise 46

(A) 19⟌19

(B) 19⟌38

(C) 19⟌57

(D) 19⟌76

(E) 19⟌95

(F) 19⟌114

(G) 19⟌133

(H) 19⟌152

(I) 19⟌171

(J) 19⟌190

(K) 19⟌209

(L) 19⟌228

(M) 19⟌247

(N) 19⟌266

(O) 19⟌285

www.math-knots.com

DP Exercise 47

(A) 20⟌20

(F) 20⟌120

(K) 20⟌220

(B) 20⟌40

(G) 20⟌140

(L) 20⟌240

(C) 20⟌60

(H) 20⟌160

(M) 20⟌260

(D) 20⟌80

(I) 20⟌180

(N) 20⟌280

(E) 20⟌100

(J) 20⟌200

(O) 20⟌300

www.math-knots.com

DP Exercise 48

(A)   21$\overline{)21}$ (F)   21$\overline{)126}$ (K)   21$\overline{)231}$

(B)   21$\overline{)42}$ (G)   21$\overline{)147}$ (L)   21$\overline{)252}$

(C)   21$\overline{)63}$ (H)   21$\overline{)168}$ (M)   21$\overline{)273}$

(D)   21$\overline{)84}$ (I)   21$\overline{)189}$ (N)   21$\overline{)294}$

(E)   21$\overline{)105}$ (J)   21$\overline{)210}$ (O)   21$\overline{)315}$

www.math-knots.com

DP Exercise 49

(A) $22\overline{)22}$

(F) $22\overline{)132}$

(K) $22\overline{)242}$

(B) $22\overline{)44}$

(G) $22\overline{)154}$

(L) $22\overline{)264}$

(C) $22\overline{)66}$

(H) $22\overline{)176}$

(M) $22\overline{)286}$

(D) $22\overline{)88}$

(I) $22\overline{)198}$

(N) $22\overline{)308}$

(E) $22\overline{)110}$

(J) $22\overline{)220}$

(O) $22\overline{)330}$

www.math-knots.com

DP Exercise 50

(A) 23)23

(B) 23)46

(C) 23)69

(D) 23)92

(E) 23)115

(F) 23)138

(G) 23)161

(H) 23)184

(I) 23)207

(J) 23)230

(K) 23)253

(L) 23)276

(M) 23)299

(N) 23)322

(O) 23)345

www.math-knots.com

DP Exercise 51

(A) $24\overline{)24}$

(F) $24\overline{)144}$

(K) $24\overline{)264}$

(B) $24\overline{)48}$

(G) $24\overline{)168}$

(L) $24\overline{)288}$

(C) $24\overline{)72}$

(H) $24\overline{)192}$

(M) $24\overline{)312}$

(D) $24\overline{)96}$

(I) $24\overline{)216}$

(N) $24\overline{)336}$

(E) $24\overline{)120}$

(J) $24\overline{)240}$

(O) $24\overline{)360}$

www.math-knots.com

DP Exercise 52

(A) 25⟌25

(B) 25⟌50

(C) 25⟌75

(D) 25⟌100

(E) 25⟌125

(F) 25⟌150

(G) 25⟌175

(H) 25⟌200

(I) 25⟌225

(J) 25⟌250

(K) 25⟌275

(L) 25⟌300

(M) 25⟌325

(N) 25⟌350

(O) 25⟌375

DP Exercise 53

(A) $50\overline{)50}$ (F) $50\overline{)300}$ (K) $50\overline{)550}$

(B) $50\overline{)100}$ (G) $50\overline{)350}$ (L) $50\overline{)600}$

(C) $50\overline{)150}$ (H) $50\overline{)400}$ (M) $50\overline{)650}$

(D) $50\overline{)200}$ (I) $50\overline{)450}$ (N) $50\overline{)700}$

(E) $50\overline{)250}$ (J) $50\overline{)500}$ (O) $50\overline{)750}$

www.math-knots.com

DP Exercise 54

(A)   100 ⟌ 50

(B)   100 ⟌ 100

(C)   100 ⟌ 150

(D)   100 ⟌ 200

(E)   100 ⟌ 250

(F)   100 ⟌ 300

(G)   100 ⟌ 350

(H)   100 ⟌ 400

(I)   100 ⟌ 450

(J)   100 ⟌ 500

(K)   100 ⟌ 550

(L)   100 ⟌ 600

(M)   100 ⟌ 650

(N)   100 ⟌ 700

(O)   100 ⟌ 750

www.math-knots.com

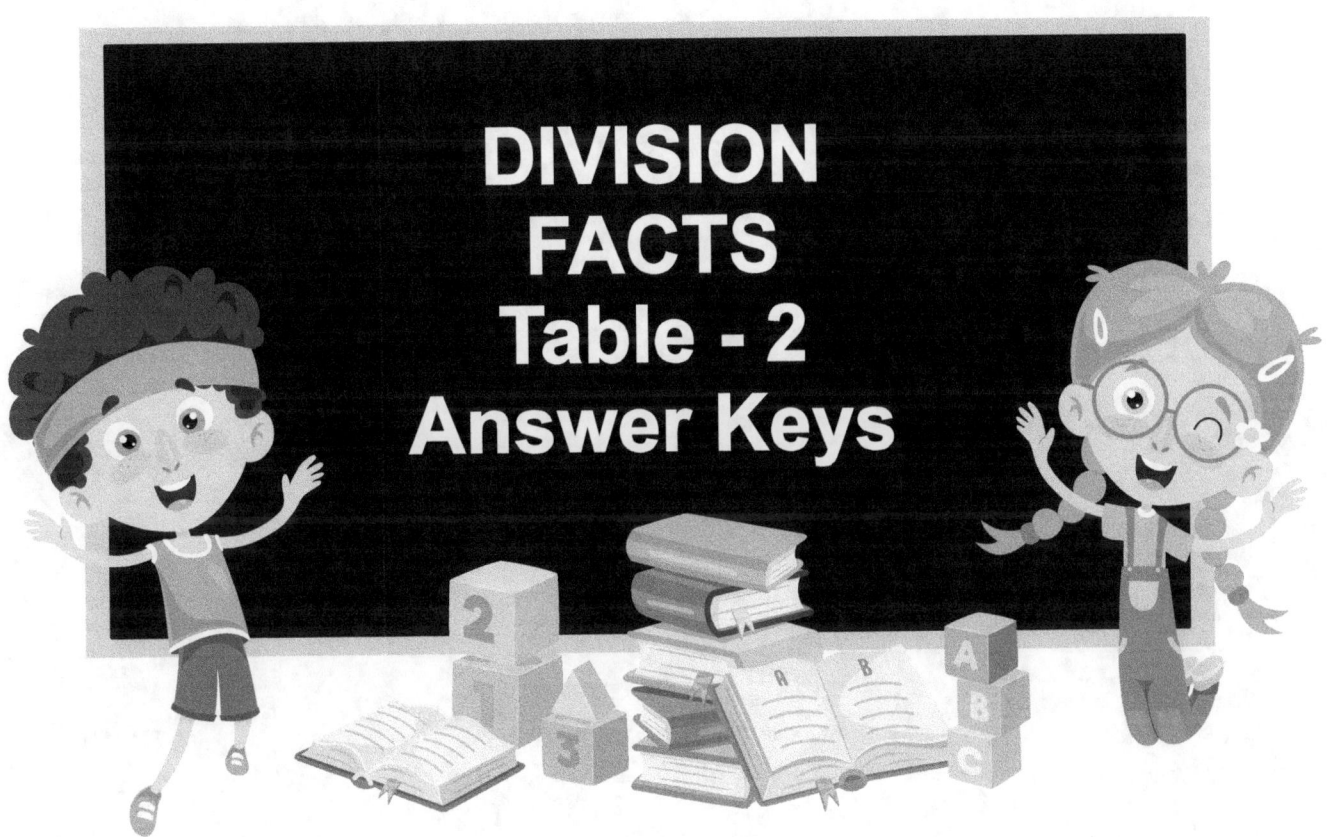

DIVISION
FACTS
Table - 2
Answer Keys

www.math-knots.com

www.math-knots.com

Exercise - 1

(A) 2⟌2

Ans : 2⟌2̄ (1)

(B) 2⟌4

Ans : 2⟌4̄ (2)

(C) 2⟌6

Ans : 2⟌6̄ (3)

(D) 2⟌8

Ans : 2⟌8̄ (4)

(E) 2⟌10

Ans : 2⟌10̄ (5)

(F) 2⟌12

Ans : 2⟌12̄ (6)

(G) 2⟌14

Ans : 2⟌14̄ (7)

(H) 2⟌16

Ans : 2⟌16̄ (8)

(I) 2⟌18

Ans : 2⟌18̄ (9)

(J) 2⟌20

Ans : 2⟌20̄ (10)

(K) 2⟌22

Ans : 2⟌22̄ (11)

(L) 2⟌24

Ans : 2⟌24̄ (12)

(M) 2⟌26

Ans : 2⟌26̄ (13)

(N) 2⟌28

Ans : 2⟌28̄ (14)

(O) 2⟌30

Ans : 2⟌30̄ (15)

Exercise - 2

1.	2 ÷ 2 =	1
2.	4 ÷ 2 =	2
3.	6 ÷ 2 =	3
4.	8 ÷ 2 =	4
5.	10 ÷ 2 =	5
6.	12 ÷ 2 =	6
7.	14 ÷ 2 =	7
8.	16 ÷ 2 =	8
9.	18 ÷ 2 =	9
10.	20 ÷ 2 =	10
11.	22 ÷ 2 =	11
12.	24 ÷ 2 =	12

1	× 2 =	2
2	× 2 =	4
3	× 2 =	6
4	× 2 =	8
5	× 2 =	10
6	× 2 =	12
7	× 2 =	14
8	× 2 =	16
9	× 2 =	18
10	× 2 =	20
11	× 2 =	22
12	× 2 =	24

Did you know division is splitting a number up by any give number.

www.math-knots.com

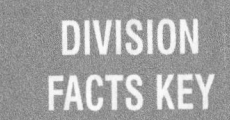

Exercise - 3

1.   C

2.   D

3.   D

4.   A

5.   C

6.   B

7.   D

8.   C

9.   A

10. B

11. D

12. B

13. C

14. A

15. D

www.math-knots.com

DIVISION FACTS KEY

Exercise - 4

Solve the maze run below.

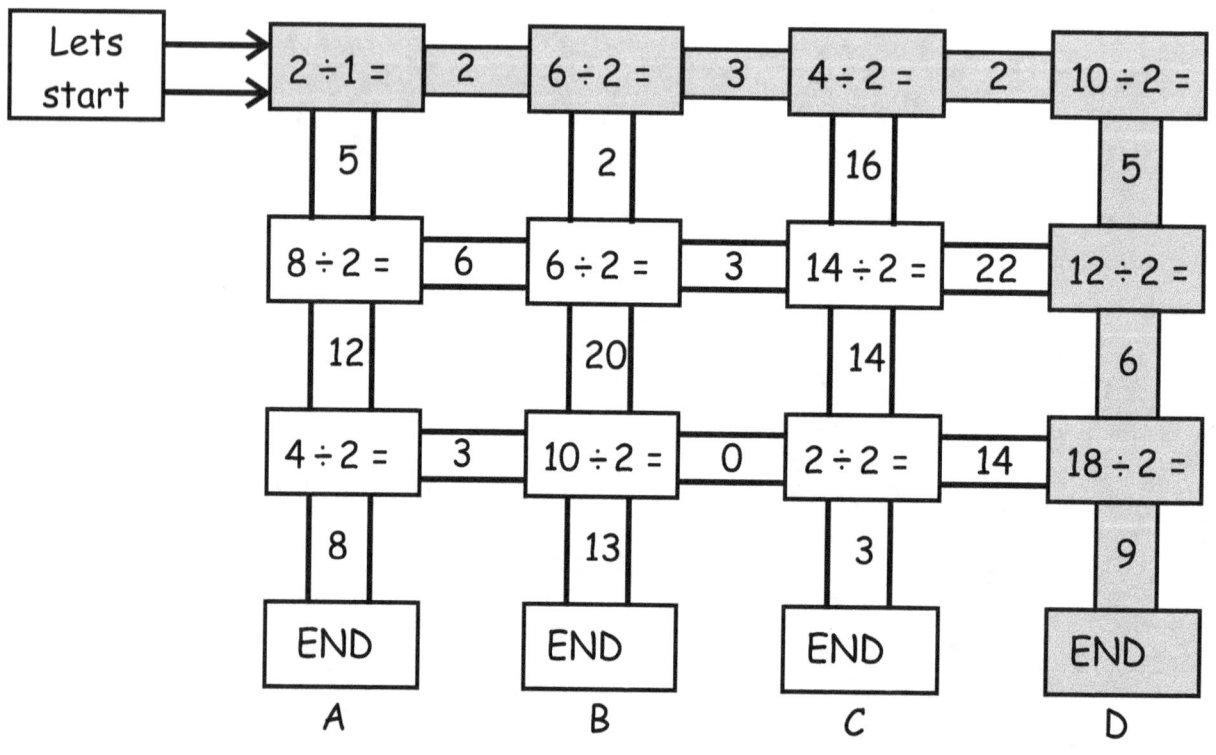

Lets start	→	2 ÷ 1 =	2	6 ÷ 2 =	3	4 ÷ 2 =	2	10 ÷ 2 =

5   2   16   5

8 ÷ 2 =   6   6 ÷ 2 =   3   14 ÷ 2 =   22   12 ÷ 2 =

12   20   14   6

4 ÷ 2 =   3   10 ÷ 2 =   0   2 ÷ 2 =   14   18 ÷ 2 =

8   13   3   9

END	END	END	END
A	B	C	D

Who won the race ? _____

www.math-knots.com

Exercise - 5

1. 2 ÷ ☐ = 1   then ☐ = ___1___

2. 4 ÷ ☐ = 2   then ☐ = ___2___

3. 6 ÷ ☐ = 2   then ☐ = ___3___

4. 8 ÷ ☐ = 2   then ☐ = ___4___

5. 10 ÷ ☐ = 2   then ☐ = ___5___

6. 12 ÷ ☐ = 2   then ☐ = ___6___

7. 14 ÷ ☐ = 2   then ☐ = ___7___

8. 16 ÷ ☐ = 2   then ☐ = ___8___

9. 18 ÷ ☐ = 2   then ☐ = ___9___

10. 20 ÷ ☐ = 2   then ☐ = ___10___

11. 22 ÷ ☐ = 2   then ☐ = ___11___

12. 24 ÷ ☐ = 2   then ☐ = ___12___

Hey you are an expert of division facts of 2!!!

www.math-knots.com

www.math-knots.com

DIVISION
FACTS
Table - 3
Answer Keys

www.math-knots.com

Exercise - 1

(A) $3\overline{)3}$

Ans : $3\overline{)3}^{\,1}$

(B) $3\overline{)6}$

Ans : $3\overline{)6}^{\,2}$

(C) $3\overline{)9}$

Ans : $3\overline{)9}^{\,3}$

(D) $3\overline{)12}$

Ans : $3\overline{)12}^{\,4}$

(E) $3\overline{)15}$

Ans : $3\overline{)15}^{\,5}$

(F) $3\overline{)18}$

Ans : $3\overline{)18}^{\,6}$

(G) $3\overline{)21}$

Ans : $3\overline{)21}^{\,7}$

(H) $3\overline{)24}$

Ans : $3\overline{)24}^{\,8}$

(I) $3\overline{)27}$

Ans : $3\overline{)27}^{\,9}$

(J) $3\overline{)30}$

Ans : $3\overline{)30}^{\,10}$

(K) $3\overline{)33}$

Ans : $3\overline{)33}^{\,11}$

(L) $3\overline{)36}$

Ans : $3\overline{)36}^{\,12}$

(M) $3\overline{)39}$

Ans : $3\overline{)39}^{\,13}$

(N) $3\overline{)42}$

Ans : $3\overline{)42}^{\,14}$

(O) $3\overline{)45}$

Ans : $3\overline{)45}^{\,15}$

www.math-knots.com

Exercise - 2

1.	3	÷ 3 =	1
2.	6	÷ 3 =	2
3.	9	÷ 3 =	3
4.	12	÷ 3 =	4
5.	15	÷ 3 =	5
6.	18	÷ 3 =	6
7.	21	÷ 3 =	7
8.	24	÷ 3 =	8
9.	27	÷ 3 =	9
10.	30	÷ 3 =	10
11.	33	÷ 3 =	11
12.	36	÷ 3 =	12

1	× 3	=	3
2	× 3	=	6
3	× 3	=	9
4	× 3	=	12
5	× 3	=	15
6	× 3	=	18
7	× 3	=	21
8	× 3	=	24
9	× 3	=	27
10	× 3	=	30
11	× 3	=	33
12	× 3	=	36

Did You Know...?

Did you know division is splitting a number up by any give number.

www.math-knots.com

 **Exercise - 3**

1. D

2. D

3. A

4. C

5. B

6. C

7. A

8. B

9. D

10. C

11. B

12. A

13. C

14. A

15. D

www.math-knots.com

Exercise - 4

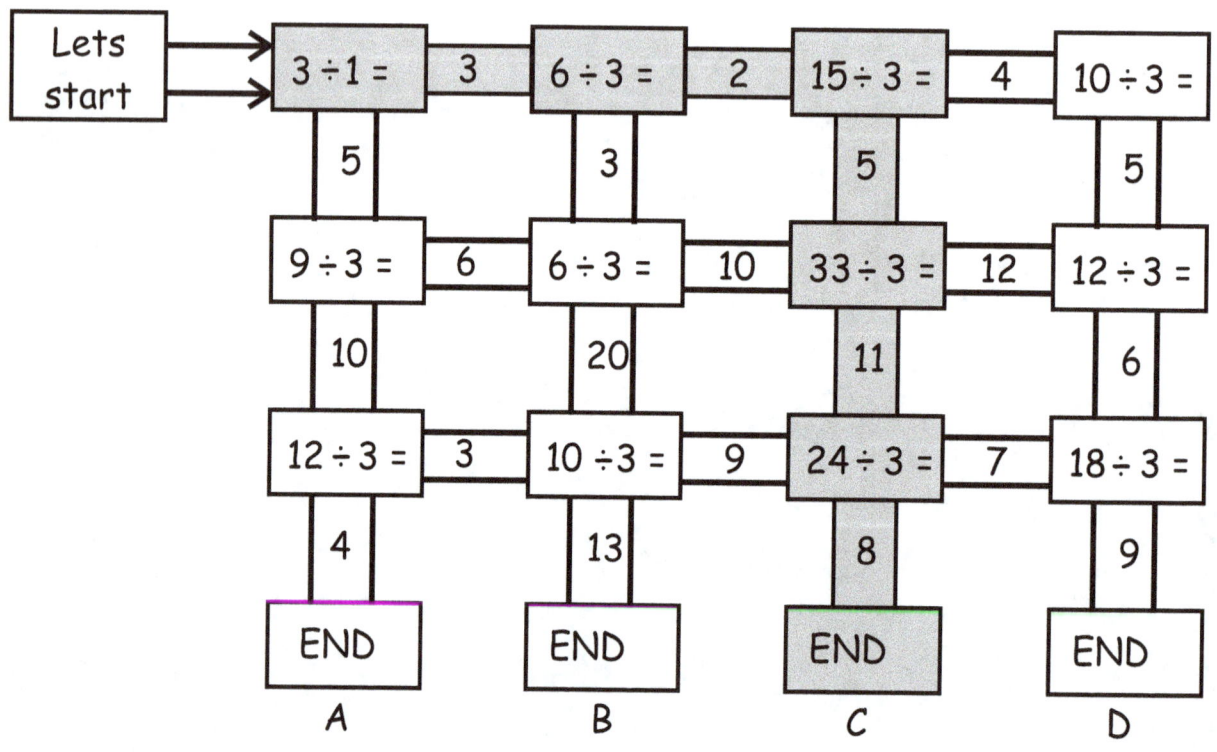

| Lets start | → | 3 ÷ 1 = | 3 | 6 ÷ 3 = | 2 | 15 ÷ 3 = | 4 | 10 ÷ 3 = |

```
Lets      →   3÷1 =   3   6÷3 =   2   15÷3 =   4   10÷3 =
start     →
                5           3           5           5
              9÷3 =   6   6÷3 =  10   33÷3 =  12   12÷3 =
               10          20          11           6
             12÷3 =   3   10÷3 =   9   24÷3 =   7   18÷3 =
                4           13          8            9
              END         END         END         END
               A           B           C           D
```

Who won the race ? ____C____

www.math-knots.com

Exercise - 5

1.   3 ÷ ☐ = 1   then   ☐ = __1__

2.   6 ÷ ☐ = 3   then   ☐ = __2__

3.   9 ÷ ☐ = 3   then   ☐ = __3__

4.   12 ÷ ☐ = 3   then   ☐ = __4__

5.   15 ÷ ☐ = 3   then   ☐ = __5__

6.   18 ÷ ☐ = 3   then   ☐ = __6__

7.   21 ÷ ☐ = 3   then   ☐ = __7__

8.   24 ÷ ☐ = 3   then   ☐ = __8__

9.   27 ÷ ☐ = 3   then   ☐ = __9__

10. 30 ÷ ☐ = 3   then   ☐ = __10__

11. 33 ÷ ☐ = 3   then   ☐ = __11__

12. 36 ÷ ☐ = 3   then   ☐ = __12__

Hey you are an expert of division facts of #3 !!!

www.math-knots.com

MULTIPLICATION
TABLE
Table - 4
Answer Keys

www.math-knots.com

Exercise - 1

(A) 4⟌4

Ans : 4⟌$\overset{1}{4}$

(B) 4⟌8

Ans : 4⟌$\overset{2}{8}$

(C) 4⟌12

Ans : 4⟌$\overset{3}{12}$

(D) 4⟌16

Ans : 4⟌$\overset{4}{16}$

(E) 4⟌20

Ans : 4⟌$\overset{5}{20}$

(F) 4⟌24

Ans : 4⟌$\overset{6}{24}$

(G) 4⟌28

Ans : 4⟌$\overset{7}{28}$

(H) 4⟌32

Ans : 4⟌$\overset{8}{32}$

(I) 4⟌36

Ans : 4⟌$\overset{9}{36}$

(J) 4⟌40

Ans : 4⟌$\overset{10}{40}$

(K) 4⟌44

Ans : 4⟌$\overset{11}{44}$

(L) 4⟌48

Ans : 4⟌$\overset{12}{48}$

(M) 4⟌52

Ans : 4⟌$\overset{13}{52}$

(N) 4⟌56

Ans : 4⟌$\overset{14}{56}$

(O) 4⟌60

Ans : 4⟌$\overset{15}{60}$

www.math-knots.com

Exercise - 2

1.	4	÷	4	=	1
2.	8	÷	4	=	2
3.	12	÷	4	=	3
4.	16	÷	4	=	4
5.	20	÷	4	=	5
6.	24	÷	4	=	6
7.	28	÷	4	=	7
8.	32	÷	4	=	8
9.	36	÷	4	=	9
10.	40	÷	4	=	10
11.	44	÷	4	=	11
12.	48	÷	4	=	12

1	×	4	=	4
2	×	4	=	8
3	×	4	=	12
4	×	4	=	16
5	×	4	=	20
6	×	4	=	24
7	×	4	=	28
8	×	4	=	32
9	×	4	=	36
10	×	4	=	40
11	×	4	=	44
12	×	4	=	48

Did you know division is splitting a number up by any give number.

www.math-knots.com

 **Exercise - 3**

1. D

2. B

3. B

4. C

5. B

6. C

7. A

8. D

9. D

10. A

11. B

12. A

13. C

14. C

15. D

www.math-knots.com

Exercise - 4

Solve the maze run below.

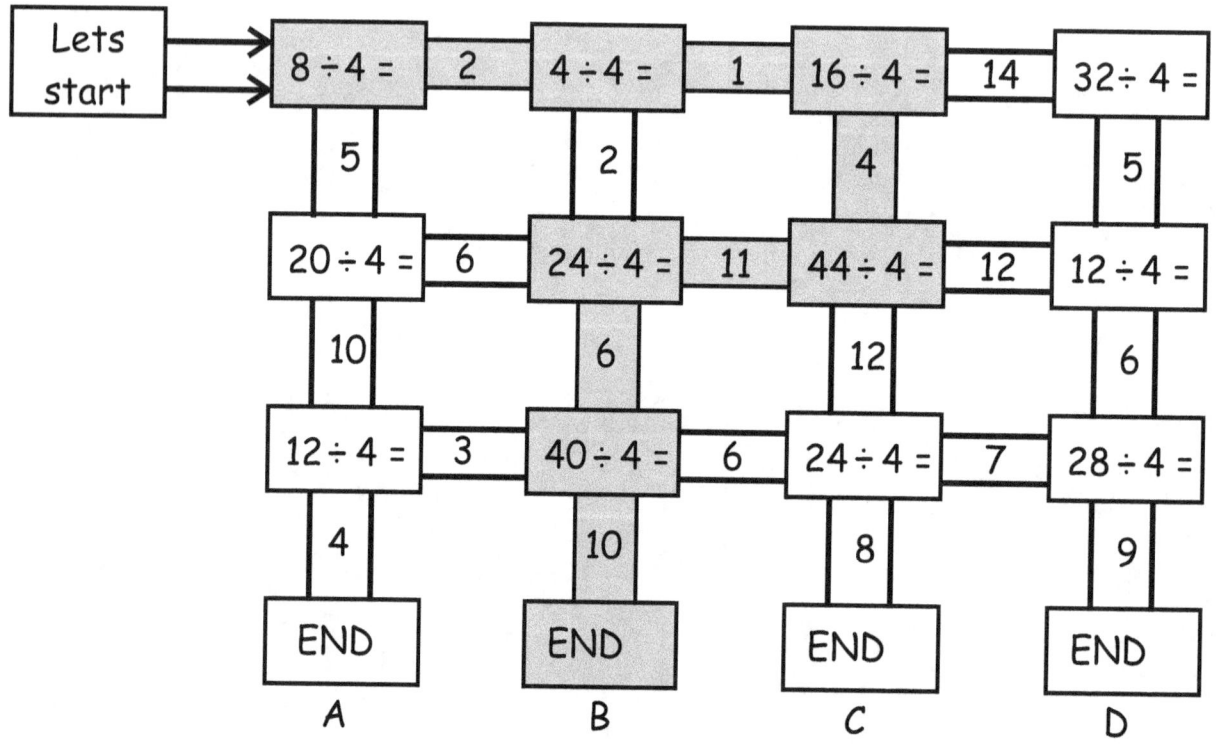

Lets start →	8 ÷ 4 =	2	4 ÷ 4 =	1	16 ÷ 4 =	14	32 ÷ 4 =

5   2   4   5

20 ÷ 4 = 6 24 ÷ 4 = 11 44 ÷ 4 = 12 12 ÷ 4 =

10   6   12   6

12 ÷ 4 = 3 40 ÷ 4 = 6 24 ÷ 4 = 7 28 ÷ 4 =

4   10   8   9

END   END   END   END
A   B   C   D

Who won the race ? _____B_____

www.math-knots.com

Exercise - 5

1. $4 \div \square = 1$ then $\square =$ ___1___

2. $8 \div \square = 4$ then $\square =$ ___2___

3. $12 \div \square = 4$ then $\square =$ ___3___

4. $16 \div \square = 4$ then $\square =$ ___4___

5. $20 \div \square = 4$ then $\square =$ ___5___

6. $24 \div \square = 4$ then $\square =$ ___6___

7. $28 \div \square = 4$ then $\square =$ ___7___

8. $32 \div \square = 4$ then $\square =$ ___8___

9. $36 \div \square = 4$ then $\square =$ ___9___

10. $40 \div \square = 4$ then $\square =$ ___10___

11. $44 \div \square = 4$ then $\square =$ ___11___

12. $48 \div \square = 4$ then $\square =$ ___12___

Hey you are an expert of division facts of #4 !!!

DIVISION
FACTS
Table - 5
Answer Keys

www.math-knots.com

www.math-knots.com

Exercise - 1

(A) $5\overline{)5}$

Ans : $5\overline{)\overset{1}{5}}$

(B) $5\overline{)10}$

Ans : $5\overline{)\overset{2}{10}}$

(C) $5\overline{)15}$

Ans : $5\overline{)\overset{3}{15}}$

(D) $5\overline{)20}$

Ans : $5\overline{)\overset{4}{20}}$

(E) $5\overline{)25}$

Ans : $5\overline{)\overset{5}{25}}$

(F) $5\overline{)30}$

Ans : $5\overline{)\overset{6}{30}}$

(G) $5\overline{)35}$

Ans : $5\overline{)\overset{7}{35}}$

(H) $5\overline{)40}$

Ans : $5\overline{)\overset{8}{40}}$

(I) $5\overline{)45}$

Ans : $5\overline{)\overset{9}{45}}$

(J) $5\overline{)50}$

Ans : $5\overline{)\overset{10}{50}}$

(K) $5\overline{)55}$

Ans : $5\overline{)\overset{11}{55}}$

(L) $5\overline{)60}$

Ans : $5\overline{)\overset{12}{60}}$

(M) $5\overline{)65}$

Ans : $5\overline{)\overset{13}{65}}$

(N) $5\overline{)70}$

Ans : $5\overline{)\overset{14}{70}}$

(O) $5\overline{)75}$

Ans : $5\overline{)\overset{15}{75}}$

www.math-knots.com

Exercise - 2

1.	5 ÷ 5	=	1	
2.	10 ÷ 5	=	2	
3.	15 ÷ 5	=	3	
4.	20 ÷ 5	=	4	
5.	25 ÷ 5	=	5	
6.	30 ÷ 5	=	6	
7.	35 ÷ 5	=	7	
8.	40 ÷ 5	=	8	
9.	45 ÷ 5	=	9	
10.	50 ÷ 5	=	10	
11.	55 ÷ 5	=	11	
12.	60 ÷ 5	=	12	

1	×	5	=	5
2	×	5	=	10
3	×	5	=	15
4	×	5	=	20
5	×	5	=	25
6	×	5	=	30
7	×	5	=	35
8	×	5	–	40
9	×	5	=	45
10	×	5	=	50
11	×	5	=	55
12	×	5	=	60

Did You Know...?

Did you know division is splitting a number up by any give number.

www.math-knots.com

 <u>**Exercise - 3**</u>

1. D

2. B

3. B

4. C

5. A

6. C

7. A

8. D

9. D

10. C

11. B

12. A

13. D

14. C

15. D

www.math-knots.com

Exercise - 4

Solve the maze run below.

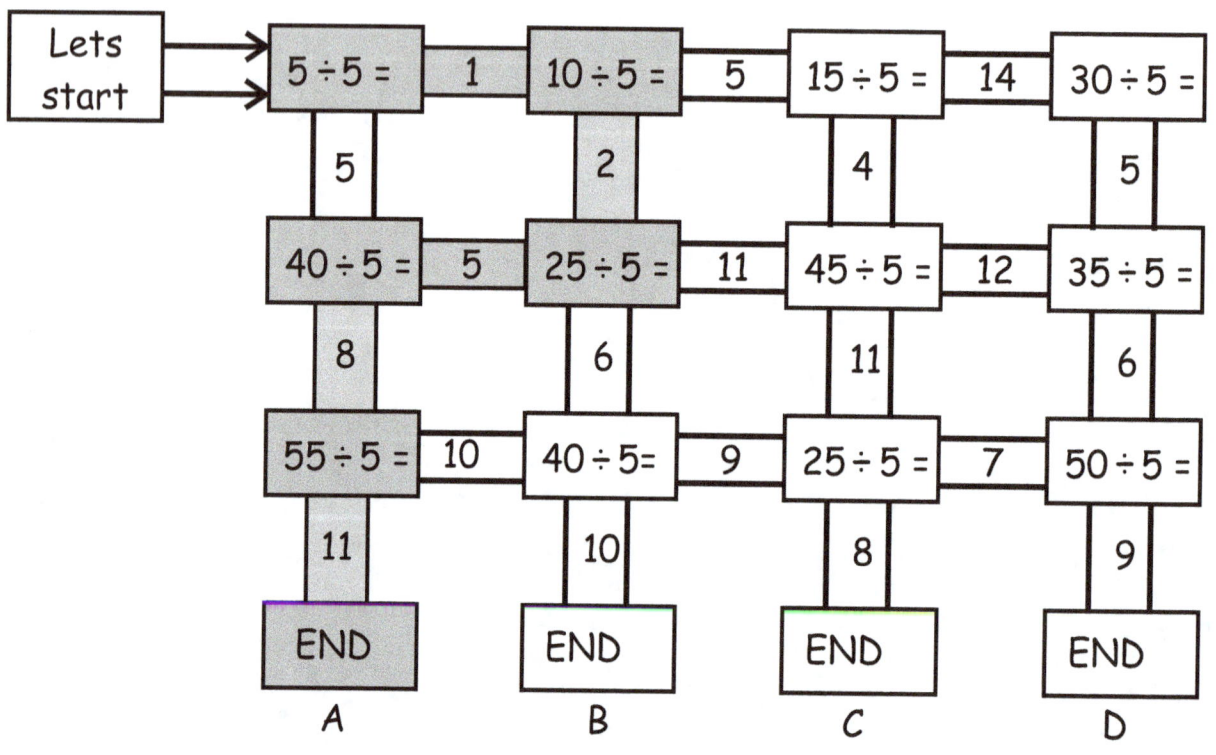

Who won the race ? _____ A _____

www.math-knots.com

Exercise - 5

1. $5 \div \boxed{} = 1$ then $\boxed{} = \underline{\quad 1 \quad}$

2. $10 \div \boxed{} = 5$ then $\boxed{} = \underline{\quad 2 \quad}$

3. $15 \div \boxed{} = 5$ then $\boxed{} = \underline{\quad 3 \quad}$

4. $20 \div \boxed{} = 5$ then $\boxed{} = \underline{\quad 4 \quad}$

5. $25 \div \boxed{} = 5$ then $\boxed{} = \underline{\quad 5 \quad}$

6. $30 \div \boxed{} = 5$ then $\boxed{} = \underline{\quad 6 \quad}$

7. $35 \div \boxed{} = 5$ then $\boxed{} = \underline{\quad 7 \quad}$

8. $40 \div \boxed{} = 5$ then $\boxed{} = \underline{\quad 8 \quad}$

9. $45 \div \boxed{} = 5$ then $\boxed{} = \underline{\quad 9 \quad}$

10. $50 \div \boxed{} = 5$ then $\boxed{} = \underline{\quad 10 \quad}$

11. $55 \div \boxed{} = 5$ then $\boxed{} = \underline{\quad 11 \quad}$

12. $60 \div \boxed{} = 5$ then $\boxed{} = \underline{\quad 12 \quad}$

Hey you are an expert of division facts of #5 !!!

www.math-knots.com

MULTIPLICATION TABLE
Table - 6
Answer Keys

www.math-knots.com

Exercise - 1

(A) 6⟌6

Ans : 6⟌6̄ (1)

(B) 6⟌12

Ans : 6⟌1̄2̄ (2)

(C) 6⟌18

Ans : 6⟌1̄8̄ (3)

(D) 6⟌24

Ans : 6⟌2̄4̄ (4)

(E) 6⟌30

Ans : 6⟌3̄0̄ (5)

(F) 6⟌36

Ans : 6⟌3̄6̄ (6)

(G) 6⟌42

Ans : 6⟌4̄2̄ (7)

(H) 6⟌48

Ans : 6⟌4̄8̄ (8)

(I) 6⟌54

Ans : 6⟌5̄4̄ (9)

(J) 6⟌60

Ans : 6⟌6̄0̄ (10)

(K) 6⟌66

Ans : 6⟌6̄6̄ (11)

(L) 6⟌72

Ans : 6⟌7̄2̄ (12)

(M) 6⟌78

Ans : 6⟌7̄8̄ (13)

(N) 6⟌84

Ans : 6⟌8̄4̄ (14)

(O) 6⟌90

Ans : 6⟌9̄0̄ (15)

www.math-knots.com

Exercise - 2

1.	$6 \div 6 =$	1
2.	$12 \div 6 =$	2
3.	$18 \div 6 =$	3
4.	$24 \div 6 =$	4
5.	$30 \div 6 =$	5
6.	$36 \div 6 =$	6
7.	$42 \div 6 =$	7
8.	$48 \div 6 =$	8
9.	$54 \div 6 =$	9
10.	$60 \div 6 =$	10
11.	$66 \div 6 =$	11
12.	$72 \div 6 =$	12

1	$\times\ 6 =$	6
2	$\times\ 6 =$	12
3	$\times\ 6 =$	18
4	$\times\ 6 =$	24
5	$\times\ 6 =$	30
6	$\times\ 6 =$	36
7	$\times\ 6 =$	42
8	$\times\ 6 =$	48
9	$\times\ 6 =$	54
10	$\times\ 6 =$	60
11	$\times\ 6 =$	66
12	$\times\ 6 =$	72

Did you know division is splitting a number up by any give number.

 **Exercise - 3**

1.   D

2.   C

3.   B

4.   C

5.   A

6.   C

7.   B

8.   A

9.   D

10.   D

11.   B

12.   A

13.   C

14.   D

15.   C

         www.math-knots.com

Exercise - 4

Solve the maze run below.

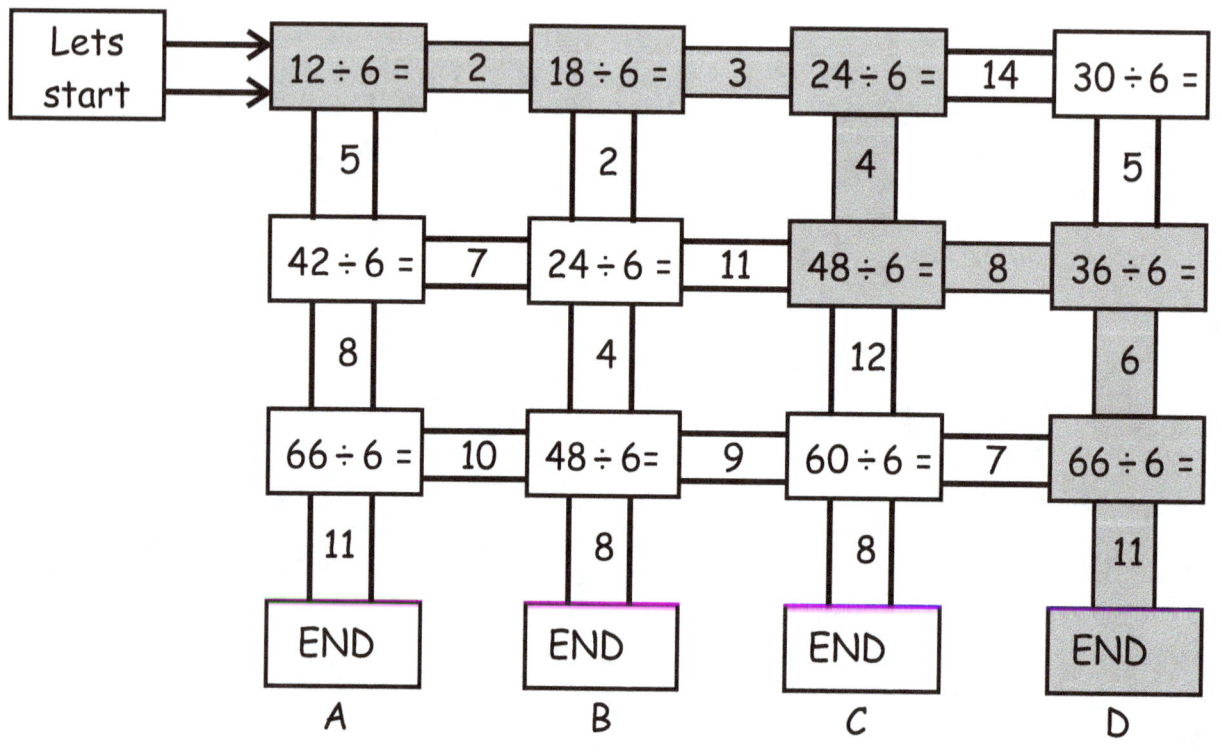

Lets start → →	12 ÷ 6 =	2	18 ÷ 6 =	3	24 ÷ 6 =	14	30 ÷ 6 =
	5		2		4		5
	42 ÷ 6 =	7	24 ÷ 6 =	11	48 ÷ 6 =	8	36 ÷ 6 =
	8		4		12		6
	66 ÷ 6 =	10	48 ÷ 6=	9	60 ÷ 6 =	7	66 ÷ 6 =
	11		8		8		11
	END A		END B		END C		END D

Who won the race ? _____D_____

www.math-knots.com

Exercise - 5

1. 6 ÷ ☐ = 1 then ☐ = ___1___

2. 12 ÷ ☐ = 6 then ☐ = ___2___

3. 18 ÷ ☐ = 6 then ☐ = ___3___

4. 24 ÷ ☐ = 6 then ☐ = ___4___

5. 30 ÷ ☐ = 6 then ☐ = ___5___

6. 36 ÷ ☐ = 6 then ☐ = ___6___

7. 42 ÷ ☐ = 6 then ☐ = ___7___

8. 48 ÷ ☐ = 6 then ☐ = ___8___

9. 54 ÷ ☐ = 6 then ☐ = ___9___

10. 60 ÷ ☐ = 6 then ☐ = ___10___

11. 66 ÷ ☐ = 6 then ☐ = ___11___

12. 72 ÷ ☐ = 6 then ☐ = ___12___

Hey you are an expert of division facts of #6 !!!

www.math-knots.com

DIVISION
FACTS
Table - 7
Answer Keys

www.math-knots.com

www.math-knots.com

Exercise - 1

(A) 7⟌7

Ans : 7⟌7 (¹)

(B) 7⟌14

Ans : 7⟌14 (²)

(C) 7⟌21

Ans : 7⟌21 (³)

(D) 7⟌28

Ans : 7⟌28 (⁴)

(E) 7⟌35

Ans : 7⟌35 (⁵)

(F) 7⟌42

Ans : 7⟌42 (⁶)

(G) 7⟌49

Ans : 7⟌49 (⁷)

(H) 7⟌56

Ans : 7⟌56 (⁸)

(I) 7⟌63

Ans : 7⟌63 (⁹)

(J) 7⟌70

Ans : 7⟌70 (¹⁰)

(K) 7⟌77

Ans : 7⟌77 (¹¹)

(L) 7⟌84

Ans : 7⟌84 (¹²)

(M) 7⟌91

Ans : 7⟌91 (¹³)

(N) 7⟌98

Ans : 7⟌98 (¹⁴)

(O) 7⟌105

Ans : 7⟌105 (¹⁵)

www.math-knots.com

Exercise - 2

1.	$7 \div 7 =$	1
2.	$14 \div 7 =$	2
3.	$21 \div 7 =$	3
4.	$28 \div 7 =$	4
5.	$35 \div 7 =$	5
6.	$42 \div 7 =$	6
7.	$49 \div 7 =$	7
8.	$56 \div 7 =$	8
9.	$63 \div 7 =$	9
10.	$70 \div 7 =$	10
11.	$77 \div 7 =$	11
12.	$84 \div 7 =$	12

1	$\times\ 7 =$	7
2	$\times\ 7 =$	14
3	$\times\ 7 =$	21
4	$\times\ 7 =$	28
5	$\times\ 7 =$	35
6	$\times\ 7 =$	42
7	$\times\ 7 =$	49
8	$\times\ 7 =$	56
9	$\times\ 7 =$	63
10	$\times\ 7 =$	70
11	$\times\ 7 =$	77
12	$\times\ 7 =$	84

Did you know division is splitting a number up by any give number.

www.math-knots.com

 <u>**Exercise - 3**</u>

1. D

2. B

3. D

4. D

5. A

6. C

7. B

8. A

9. D

10. D

11. A

12. C

13. B

14. A

15. A

www.math-knots.com

Exercise - 4

Solve the maze run below.

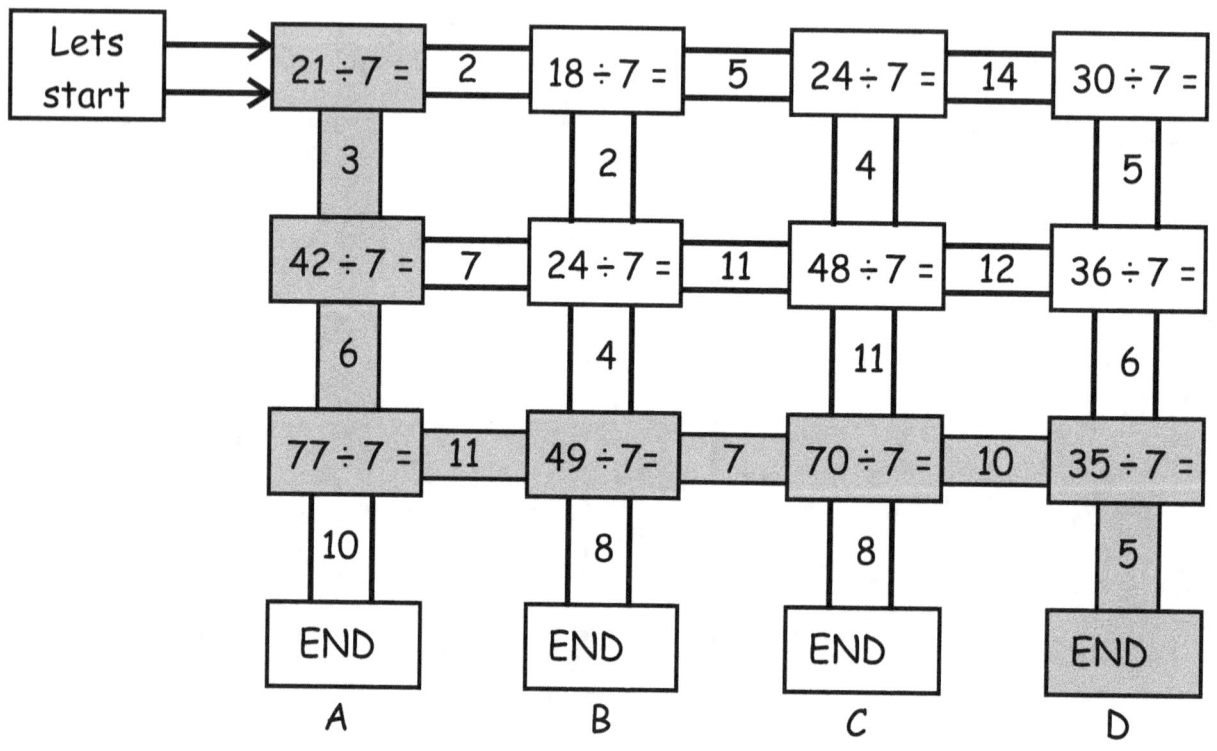

Lets start	→ 21 ÷ 7 =	2	18 ÷ 7 =	5	24 ÷ 7 =	14	30 ÷ 7 =
	3		2		4		5
	42 ÷ 7 =	7	24 ÷ 7 =	11	48 ÷ 7 =	12	36 ÷ 7 =
	6		4		11		6
	77 ÷ 7 =	11	49 ÷ 7=	7	70 ÷ 7 =	10	35 ÷ 7 =
	10		8		8		5
	END		END		END		END
	A		B		C		D

Who won the race ? _____D_____

www.math-knots.com

Exercise - 5

1. 7 ÷ ☐ = 1 then ☐ = ___1___

2. 14 ÷ ☐ = 7 then ☐ = ___2___

3. 21 ÷ ☐ = 7 then ☐ = ___3___

4. 28 ÷ ☐ = 7 then ☐ = ___4___

5. 35 ÷ ☐ = 7 then ☐ = ___5___

6. 42 ÷ ☐ = 7 then ☐ = ___6___

7. 49 ÷ ☐ = 7 then ☐ = ___7___

8. 56 ÷ ☐ = 7 then ☐ = ___8___

9. 63 ÷ ☐ = 7 then ☐ = ___9___

10. 70 ÷ ☐ = 7 then ☐ = ___10___

11. 77 ÷ ☐ = 7 then ☐ = ___11___

12. 84 ÷ ☐ = 7 then ☐ = ___12___

Hey you are an expert of division facts of #7 !!!

www.math-knots.com

www.math-knots.com

DIVISION
FACTS
Table - 8
Answer Keys

www.math-knots.com

www.math-knots.com

Exercise - 1

(A) 8⟌8

Ans : 8⟌8̄ (1)

(B) 8⟌16

Ans : 8⟌16̄ (2)

(C) 8⟌24

Ans : 8⟌24̄ (3)

(D) 8⟌32

Ans : 8⟌32̄ (4)

(E) 8⟌40

Ans : 8⟌40̄ (5)

(F) 8⟌48

Ans : 8⟌48̄ (6)

(G) 8⟌56

Ans : 8⟌56̄ (7)

(H) 8⟌64

Ans : 8⟌64̄ (8)

(I) 8⟌72

Ans : 8⟌72̄ (9)

(J) 8⟌80

Ans : 8⟌80̄ (10)

(K) 8⟌88

Ans : 8⟌88̄ (11)

(L) 8⟌96

Ans : 8⟌96̄ (12)

(M) 8⟌104

Ans : 8⟌104̄ (13)

(N) 8⟌112

Ans : 8⟌112̄ (14)

(O) 8⟌120

Ans : 8⟌120̄ (15)

www.math-knots.com

Exercise - 2

1.	8	÷	8	=	1	1	×	8	= 8
2.	16	÷	8	=	2	2	×	8	= 16
3.	24	÷	8	=	3	3	×	8	= 24
4.	32	÷	8	=	4	4	×	8	= 32
5.	40	÷	8	=	5	5	×	8	= 40
6.	48	÷	8	=	6	6	×	8	= 48
7.	56	÷	8	=	7	7	×	8	= 56
8.	64	÷	8	=	8	8	×	8	= 64
9.	72	÷	8	=	9	9	×	8	= 72
10.	80	÷	8	=	10	10	×	8	= 80
11.	88	÷	8	=	11	11	×	8	= 88
12.	96	÷	8	=	12	12	×	8	= 96

Did you know division is splitting a number up by any give number.

www.math-knots.com

 **Exercise - 3**

1. D

2. C

3. A

4. D

5. B

6. D

7. A

8. C

9. C

10. D

11. A

12. D

13. B

14. B

15. C

www.math-knots.com

Exercise - 4

Solve the maze run below.

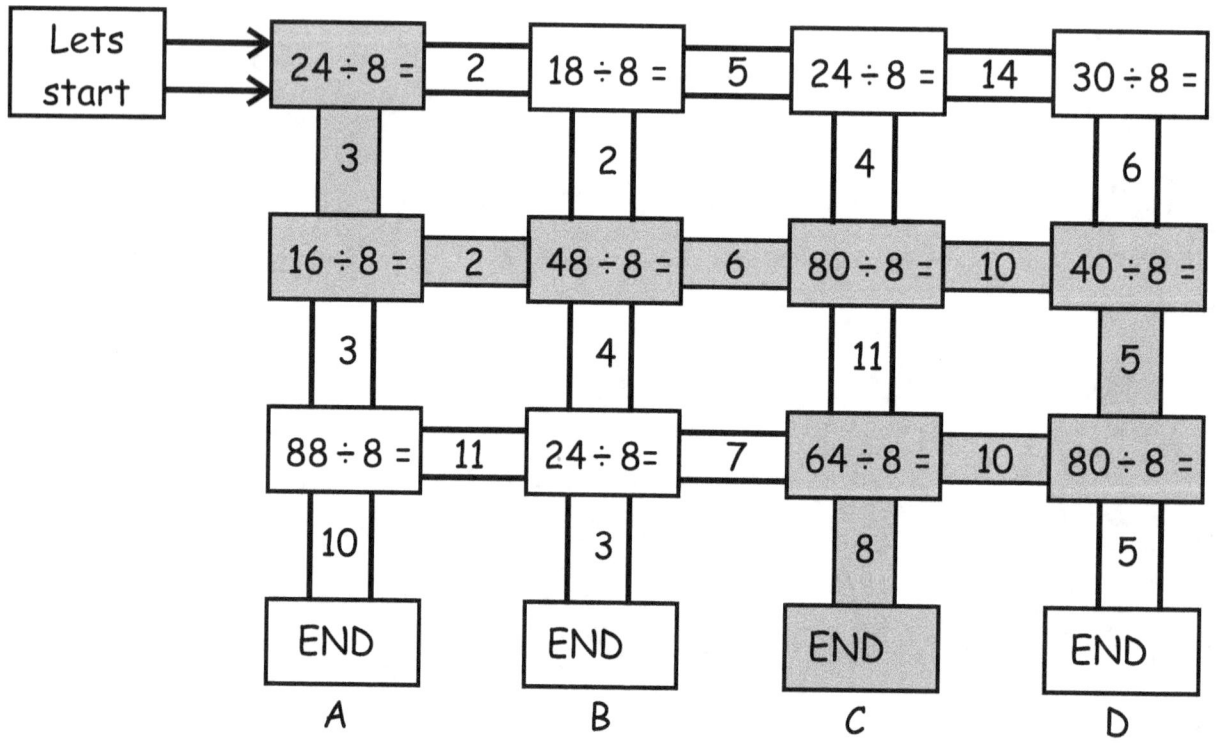

Who won the race ? _____ C _____

Exercise - 5

1. 8 ÷ ☐ = 1 then ☐ = __1__

2. 16 ÷ ☐ = 8 then ☐ = __2__

3. 24 ÷ ☐ = 8 then ☐ = __3__

4. 32 ÷ ☐ = 8 then ☐ = __4__

5. 40 ÷ ☐ = 8 then ☐ = __5__

6. 48 ÷ ☐ = 8 then ☐ = __6__

7. 56 ÷ ☐ = 8 then ☐ = __7__

8. 64 ÷ ☐ = 8 then ☐ = __8__

9. 72 ÷ ☐ = 8 then ☐ = __9__

10. 80 ÷ ☐ = 8 then ☐ = __10__

11. 88 ÷ ☐ = 8 then ☐ = __11__

12. 96 ÷ ☐ = 8 then ☐ = __12__

Hey you are an expert of division facts of #8 !!!

DIVISION
FACTS
Table - 9
Answer Keys

www.math-knots.com

Exercise - 1

(A) $9\overline{)9}$

Ans : $9\overline{)9}^{\,1}$

(B) $9\overline{)18}$

Ans : $9\overline{)18}^{\,2}$

(C) $9\overline{)27}$

Ans : $9\overline{)27}^{\,3}$

(D) $9\overline{)36}$

Ans : $9\overline{)36}^{\,4}$

(E) $9\overline{)45}$

Ans : $9\overline{)45}^{\,5}$

(F) $9\overline{)54}$

Ans : $9\overline{)54}^{\,6}$

(G) $9\overline{)63}$

Ans : $9\overline{)63}^{\,7}$

(H) $9\overline{)72}$

Ans : $9\overline{)72}^{\,8}$

(I) $9\overline{)81}$

Ans : $9\overline{)81}^{\,9}$

(J) $9\overline{)90}$

Ans : $9\overline{)90}^{\,10}$

(K) $9\overline{)99}$

Ans : $9\overline{)99}^{\,11}$

(L) $9\overline{)108}$

Ans : $9\overline{)108}^{\,12}$

(M) $9\overline{)117}$

Ans : $9\overline{)117}^{\,13}$

(N) $9\overline{)126}$

Ans : $9\overline{)126}^{\,14}$

(O) $9\overline{)135}$

Ans : $9\overline{)135}^{\,15}$

Exercise - 2

1.	9 ÷ 9 =	1
2.	18 ÷ 9 =	2
3.	27 ÷ 9 =	3
4.	36 ÷ 9 =	4
5.	45 ÷ 9 =	5
6.	54 ÷ 9 =	6
7.	63 ÷ 9 =	7
8.	72 ÷ 9 =	8
9.	81 ÷ 9 =	9
10.	90 ÷ 9 =	10
11.	99 ÷ 9 =	11
12.	108 ÷ 9 =	12

1 × 9 =	9
2 × 9 =	18
3 × 9 =	27
4 × 9 =	36
5 × 9 =	45
6 × 9 =	54
7 × 9 =	63
8 × 9 =	72
9 × 9 =	81
10 × 9 =	90
11 × 9 =	99
12 × 9 =	108

Did You Know...?

Did you know division is splitting a number up by any give number.

www.math-knots.com

 **<u>Exercise - 3</u>**

1. A

2. D

3. A

4. C

5. C

6. B

7. C

8. C

9. A

10. C

11. D

12. B

13. A

14. A

15. D

www.math-knots.com

Exercise - 4

Solve the maze run below.

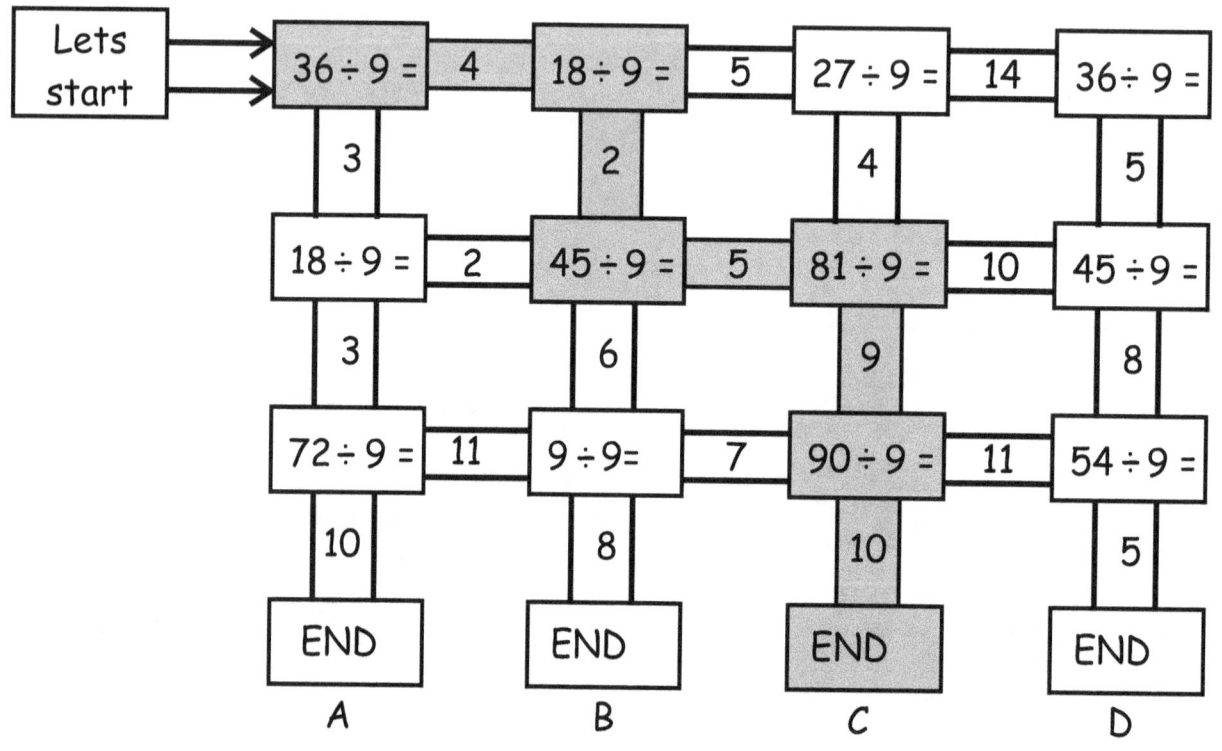

Who won the race ? _____C_____

Exercise - 5

1. 9 ÷ ☐ = 1 then ☐ = ___1___

2. 18 ÷ ☐ = 9 then ☐ = ___2___

3. 27 ÷ ☐ = 9 then ☐ = ___3___

4. 36 ÷ ☐ = 9 then ☐ = ___4___

5. 45 ÷ ☐ = 9 then ☐ = ___5___

6. 54 ÷ ☐ = 9 then ☐ = ___6___

7. 63 ÷ ☐ = 9 then ☐ = ___7___

8. 72 ÷ ☐ = 9 then ☐ = ___8___

9. 81 ÷ ☐ = 9 then ☐ = ___9___

10. 90 ÷ ☐ = 9 then ☐ = ___10___

11. 99 ÷ ☐ = 9 then ☐ = ___11___

12. 108 ÷ ☐ = 9 then ☐ = ___12___

Hey you are an expert of division facts of #9 !!!

www.math-knots.com

MULTIPLICATION TABLE
Table - 10
Answer Keys

www.math-knots.com

www.math-knots.com

Exercise - 1

(A) 10⟌10

Ans : 10⟌$\overset{1}{10}$

(B) 10⟌20

Ans : 10⟌$\overset{2}{20}$

(C) 10⟌30

Ans : 10⟌$\overset{3}{30}$

(D) 10⟌40

Ans : 10⟌$\overset{4}{40}$

(E) 10⟌50

Ans : 10⟌$\overset{5}{50}$

(F) 10⟌60

Ans : 10⟌$\overset{6}{60}$

(G) 10⟌70

Ans : 10⟌$\overset{7}{70}$

(H) 10⟌80

Ans : 10⟌$\overset{8}{80}$

(I) 10⟌90

Ans : 10⟌$\overset{9}{90}$

(J) 10⟌100

Ans : 10⟌$\overset{10}{100}$

(K) 10⟌110

Ans : 10⟌$\overset{11}{110}$

(L) 10⟌120

Ans : 10⟌$\overset{12}{120}$

(M) 10⟌130

Ans : 10⟌$\overset{13}{130}$

(N) 10⟌140

Ans : 10⟌$\overset{14}{140}$

(O) 10⟌150

Ans : 10⟌$\overset{15}{150}$

Exercise - 2

1.	$10 \div 10 =$	1
2.	$20 \div 10 =$	2
3.	$30 \div 10 =$	3
4.	$40 \div 10 =$	4
5.	$50 \div 10 =$	5
6.	$60 \div 10 =$	6
7.	$70 \div 10 =$	7
8.	$80 \div 10 =$	8
9.	$90 \div 10 =$	9
10.	$100 \div 10 =$	10
11.	$110 \div 10 =$	11
12.	$120 \div 10 =$	12

1	$\times\ 10 =$	10
2	$\times\ 10 =$	20
3	$\times\ 10 =$	30
4	$\times\ 10 =$	40
5	$\times\ 10 =$	50
6	$\times\ 10 =$	60
7	$\times\ 10 =$	70
8	$\times\ 10 =$	80
9	$\times\ 10 =$	90
10	$\times\ 10 =$	100
11	$\times\ 10 =$	110
12	$\times\ 10 =$	120

Did you know division is splitting a number up by any give number.

www.math-knots.com

 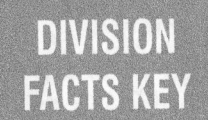

 **Exercise - 3**

1. C

2. A

3. C

4. D

5. A

6. D

7. B

8. B

9. A

10. C

11. D

12. A

13. D

14. C

15. C

 www.math-knots.com

Exercise - 4

Solve the maze run below.

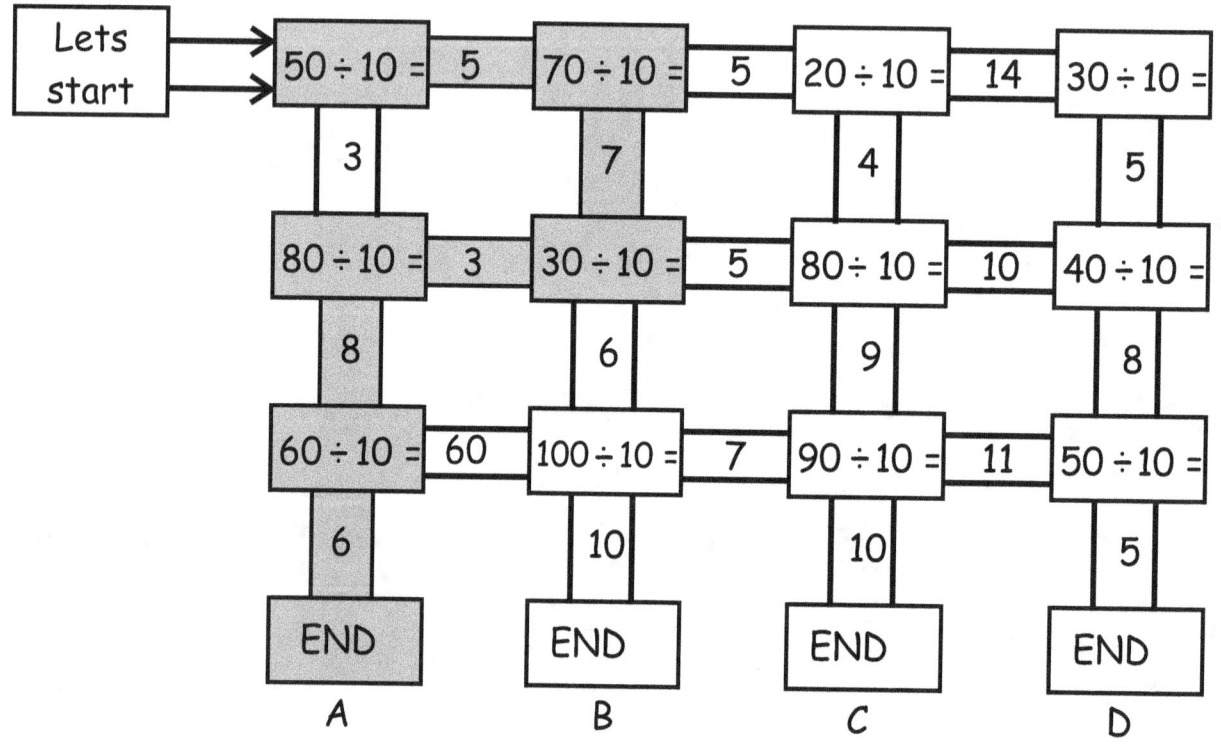

Lets start → →	50 ÷ 10 = 5	70 ÷ 10 = 5	20 ÷ 10 = 14	30 ÷ 10 =
	3	7	4	5
	80 ÷ 10 = 3	30 ÷ 10 = 5	80 ÷ 10 = 10	40 ÷ 10 =
	8	6	9	8
	60 ÷ 10 = 60	100 ÷ 10 = 7	90 ÷ 10 = 11	50 ÷ 10 =
	6	10	10	5
	END A	END B	END C	END D

Who won the race ? _____A_____

Exercise - 5

1. $10 \div \square = 1$ then $\square$ = _____1_____

2. $20 \div \square = 10$ then $\square$ = _____2_____

3. $30 \div \square = 10$ then $\square$ = _____3_____

4. $40 \div \square = 10$ then $\square$ = _____4_____

5. $50 \div \square = 10$ then $\square$ = _____5_____

6. $60 \div \square = 10$ then $\square$ = _____6_____

7. $70 \div \square = 10$ then $\square$ = _____7_____

8. $80 \div \square = 10$ then $\square$ = _____8_____

9. $90 \div \square = 10$ then $\square$ = _____9_____

10. $100 \div \square = 10$ then $\square$ = _____10_____

11. $110 \div \square = 10$ then $\square$ = _____11_____

12. $120 \div \square = 10$ then $\square$ = _____12_____

Hey you are an expert of division facts #10 !!!

www.math-knots.com

Exercise - 1

(A) 11 ⌐11

Ans : 11 ⌐$\overset{1}{11}$

(B) 11 ⌐22

Ans : 11 ⌐$\overset{2}{22}$

(C) 11 ⌐33

Ans : 11 ⌐$\overset{3}{33}$

(D) 11 ⌐44

Ans : 11 ⌐$\overset{4}{44}$

(E) 11 ⌐55

Ans : 11 ⌐$\overset{5}{55}$

(F) 11 ⌐66

Ans : 11 ⌐$\overset{6}{66}$

(G) 11 ⌐77

Ans : 11 ⌐$\overset{7}{77}$

(H) 11 ⌐88

Ans : 11 ⌐$\overset{8}{88}$

(I) 11 ⌐99

Ans : 11 ⌐$\overset{9}{99}$

(J) 11 ⌐110

Ans : 11 ⌐$\overset{10}{110}$

(K) 11 ⌐121

Ans : 11 ⌐$\overset{11}{121}$

(L) 11 ⌐132

Ans : 11 ⌐$\overset{12}{132}$

(M) 11 ⌐143

Ans : 11 ⌐$\overset{13}{143}$

(N) 11 ⌐154

Ans : 11 ⌐$\overset{14}{154}$

(O) 11 ⌐165

Ans : 11 ⌐$\overset{15}{165}$

www.math-knots.com

Exercise - 2

1.	$11 \div 11 = $ ___1___
2.	$22 \div 11 = $ ___2___
3.	$33 \div 11 = $ ___3___
4.	$44 \div 11 = $ ___4___
5.	$55 \div 11 = $ ___5___
6.	$66 \div 11 = $ ___6___
7.	$77 \div 11 = $ ___7___
8.	$88 \div 11 = $ ___8___
9.	$99 \div 11 = $ ___9___
10.	$110 \div 11 = $ ___10___
11.	$121 \div 11 = $ ___11___
12.	$132 \div 11 = $ ___12___

1	$\times$ ___11___ $= 11$
2	$\times$ ___11___ $= 22$
3	$\times$ ___11___ $= 33$
4	$\times$ ___11___ $= 44$
5	$\times$ ___11___ $= 55$
6	$\times$ ___11___ $= 66$
7	$\times$ ___11___ $= 77$
8	$\times$ ___11___ $= 88$
9	$\times$ ___11___ $= 99$
10	$\times$ ___11___ $= 110$
11	$\times$ ___11___ $= 121$
12	$\times$ ___11___ $= 132$

Did You Know...?

Did you know division is splitting a number up by any give number.

www.math-knots.com

 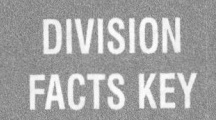

 **Exercise - 3**

1. B

2. A

3. C

4. A

5. D

6. A

7. D

8. C

9. D

10. B

11. A

12. D

13. A

14. C

15. D

www.math-knots.com

Exercise - 4

Solve the maze run below.

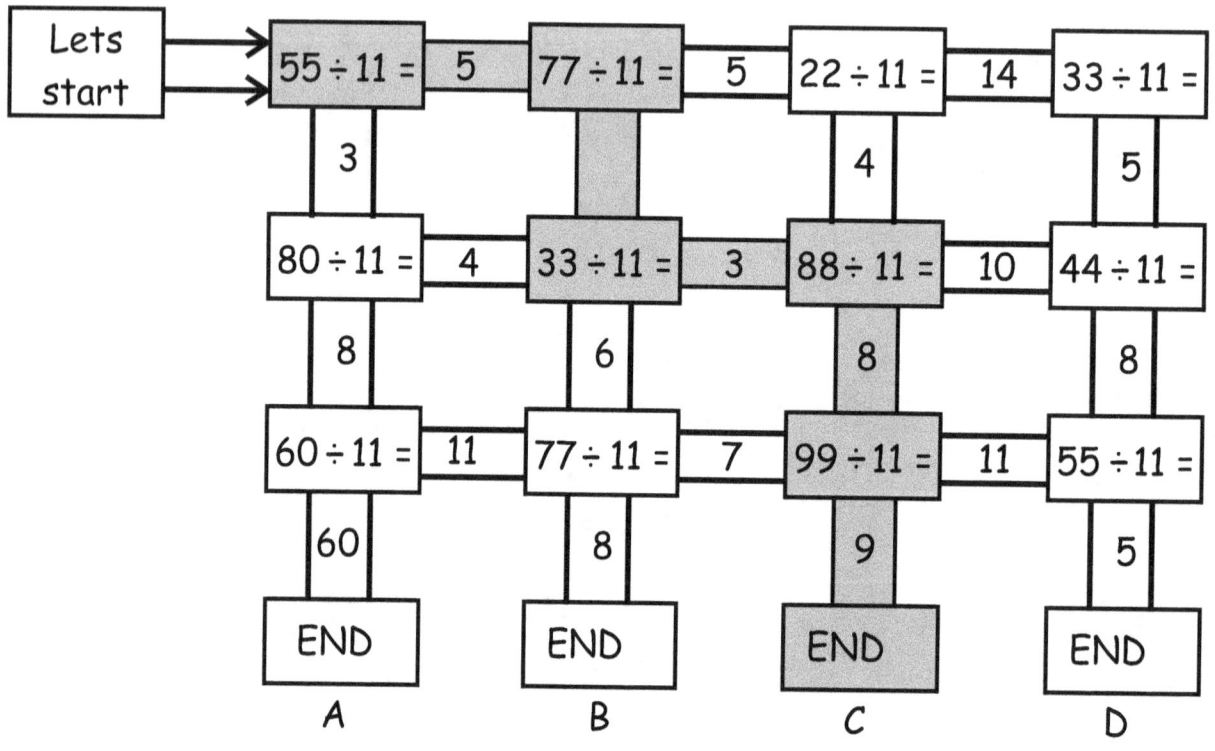

Lets start	55 ÷ 11 =	5	77 ÷ 11 =	5	22 ÷ 11 =	14	33 ÷ 11 =
	3				4		5
	80 ÷ 11 =	4	33 ÷ 11 =	3	88 ÷ 11 =	10	44 ÷ 11 =
	8		6		8		8
	60 ÷ 11 =	11	77 ÷ 11 =	7	99 ÷ 11 =	11	55 ÷ 11 =
	60		8		9		5
	END A		END B		END C		END D

Who won the race ? _____ C _____

www.math-knots.com

Exercise - 5

1. 11 ÷ ☐ = 1 then ☐ = ___1___

2. 22 ÷ ☐ = 11 then ☐ = ___2___

3. 33 ÷ ☐ = 11 then ☐ = ___3___

4. 44 ÷ ☐ = 11 then ☐ = ___4___

5. 55 ÷ ☐ = 11 then ☐ = ___5___

6. 66 ÷ ☐ = 11 then ☐ = ___6___

7. 77 ÷ ☐ = 11 then ☐ = ___7___

8. 88 ÷ ☐ = 11 then ☐ = ___8___

9. 99 ÷ ☐ = 11 then ☐ = ___9___

10. 110 ÷ ☐ = 11 then ☐ = ___10___

11. 121 ÷ ☐ = 11 then ☐ = ___11___

12. 132 ÷ ☐ = 11 then ☐ = ___12___

Hey you are an expert of division facts of #11 !!!

www.math-knots.com

DIVISION
FACTS
Table - 12
Answer Keys

www.math-knots.com

Exercise - 1

(A) 12⟌12

Ans : 12⟌12 ¹

(B) 12⟌24

Ans : 12⟌24 ²

(C) 12⟌36

Ans : 12⟌36 ³

(D) 12⟌48

Ans : 12⟌48 ⁴

(E) 12⟌60

Ans : 12⟌60 ⁵

(F) 12⟌72

Ans : 12⟌72 ⁶

(G) 12⟌84

Ans : 12⟌84 ⁷

(H) 12⟌96

Ans : 12⟌96 ⁸

(I) 12⟌108

Ans : 12⟌108 ⁹

(J) 12⟌120

Ans : 12⟌120 ¹⁰

(K) 12⟌132

Ans : 12⟌132 ¹¹

(L) 12⟌144

Ans : 12⟌144 ¹²

(M) 12⟌156

Ans : 12⟌156 ¹³

(N) 12⟌168

Ans : 12⟌168 ¹⁴

(O) 12⟌180

Ans : 12⟌180 ¹⁵

www.math-knots.com

Exercise - 2

1.	$12 \div 12 =$	1
2.	$24 \div 12 =$	2
3.	$36 \div 12 =$	3
4.	$48 \div 12 =$	4
5.	$60 \div 12 =$	5
6.	$72 \div 12 =$	6
7.	$84 \div 12 =$	7
8.	$96 \div 12 =$	8
9.	$108 \div 12 =$	9
10.	$120 \div 12 =$	10
11.	$132 \div 12 =$	11
12.	$144 \div 12 =$	12

1	$\times$	12	$= 12$
2	$\times$	12	$= 24$
3	$\times$	12	$= 36$
4	$\times$	12	$= 48$
5	$\times$	12	$= 60$
6	$\times$	12	$= 72$
7	$\times$	12	$= 84$
8	$\times$	12	$= 96$
9	$\times$	12	$= 108$
10	$\times$	12	$= 120$
11	$\times$	12	$= 132$
12	$\times$	12	$= 144$

Did you know division is splitting a number up by any give number.

www.math-knots.com

Exercise - 3

1. A

2. A

3. C

4. D

5. C

6. A

7. C

8. B

9. B

10. C

11. D

12. C

13. B

14. D

15. A

Exercise - 4

Solve the maze run below.

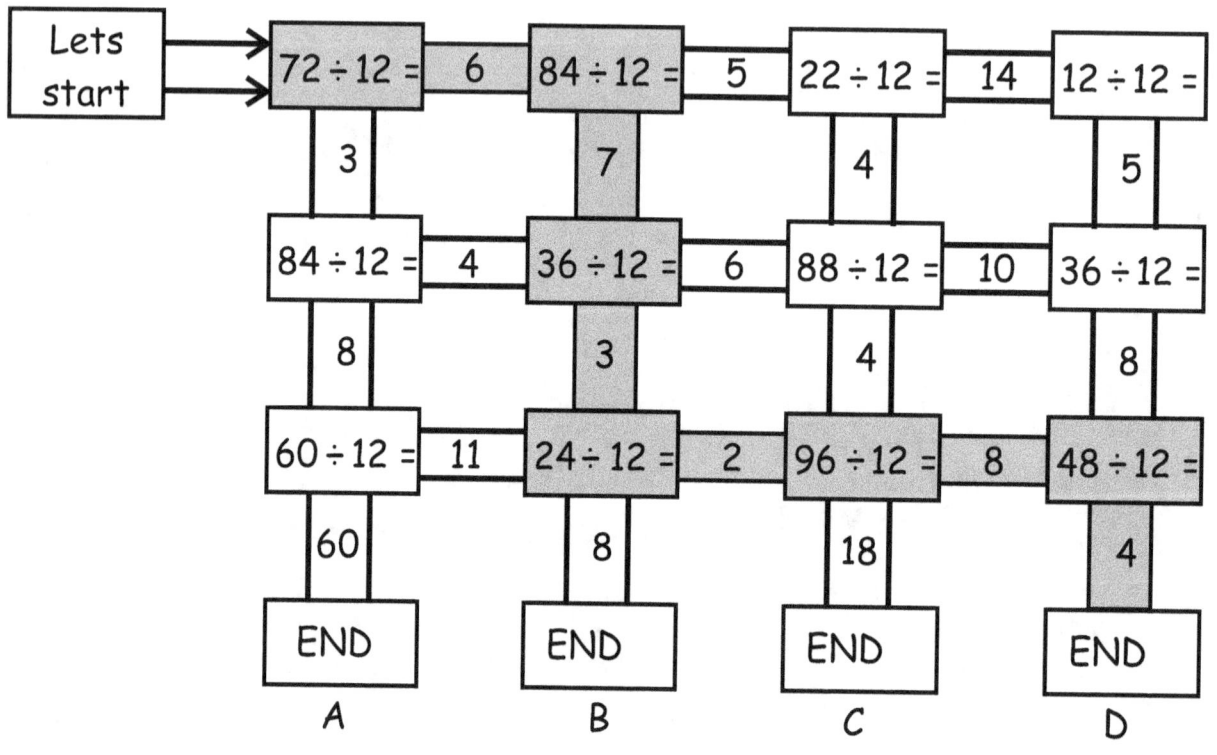

Lets start	→	72 ÷ 12 =	6	84 ÷ 12 =	5	22 ÷ 12 =	14	12 ÷ 12 =

Who won the race ? _____D_____

www.math-knots.com

Exercise - 5

1.   12 ÷ ☐ = 1   then   ☐ = <u> 1 </u>

2.   24 ÷ ☐ = 12   then   ☐ = <u> 2 </u>

3.   36 ÷ ☐ = 12   then   ☐ = <u> 3 </u>

4.   48 ÷ ☐ = 12   then   ☐ = <u> 4 </u>

5.   60 ÷ ☐ = 12   then   ☐ = <u> 5 </u>

6.   72 ÷ ☐ = 12   then   ☐ = <u> 6 </u>

7.   84 ÷ ☐ = 12   then   ☐ = <u> 7 </u>

8.   96 ÷ ☐ = 12   then   ☐ = <u> 8 </u>

9.   108 ÷ ☐ = 12   then   ☐ = <u> 9 </u>

10.   120 ÷ ☐ = 12   then   ☐ = <u> 10 </u>

11.   132 ÷ ☐ = 12   then   ☐ = <u> 11 </u>

12.   144 ÷ ☐ = 12   then   ☐ = <u> 12 </u>

Hey you are an expert of division facts of #12 !!!

www.math-knots.com

www.math-knots.com

DP Exercise 1

1. $1 \div 1 =$ ___1___

2. $2 \div 1 =$ ___2___

3. $3 \div 1 =$ ___3___

4. $4 \div 1 =$ ___4___

5. $5 \div 1 =$ ___5___

6. $6 \div 1 =$ ___6___

7. $7 \div 1 =$ ___7___

8. $8 \div 1 =$ ___8___

9. $9 \div 1 =$ ___9___

10. $10 \div 1 =$ ___10___

11. $11 \div 1 =$ ___11___

12. $12 \div 1 =$ ___12___

Did you know any number when divided by one is the number itself ?

www.math-knots.com

DP Exercise 2

1.	2 ÷ 2 =	1
2.	4 ÷ 2 =	2
3.	6 ÷ 2 =	3
4.	8 ÷ 2 =	4
5.	10 ÷ 2 =	5
6.	12 ÷ 2 =	6
7.	14 ÷ 2 =	7
8.	16 ÷ 2 =	8
9.	18 ÷ 2 =	9
10.	20 ÷ 2 =	10
11.	22 ÷ 2 =	11
12.	24 ÷ 2 =	12

1 × 2 = 2
2 × 2 = 4
3 × 2 = 6
4 × 2 = 8
5 × 2 = 10
6 × 2 = 12
7 × 2 = 14
8 × 2 = 16
9 × 2 = 18
10 × 2 = 20
11 × 2 = 22
12 × 2 = 24

Did you know division by 2 means
dividing the given number into 2 equal halfs ?

DP Exercise 3

1.	3 ÷ 3 =	1
2.	6 ÷ 3 =	2
3.	9 ÷ 3 =	3
4.	12 ÷ 3 =	4
5.	15 ÷ 3 =	5
6.	18 ÷ 3 =	6
7.	21 ÷ 3 =	7
8.	24 ÷ 3 =	8
9.	27 ÷ 3 =	9
10.	30 ÷ 3 =	10
11.	33 ÷ 3 =	11
12.	36 ÷ 3 =	12

1	× 3 =	3
2	× 3 =	6
3	× 3 =	9
4	× 3 =	12
5	× 3 =	15
6	× 3 =	18
7	× 3 =	21
8	× 3 =	24
9	× 3 =	27
10	× 3 =	30
11	× 3 =	33
12	× 3 =	36

Did you know division by 3 means
dividing the given number into 3 equal halfs ?

www.math-knots.com

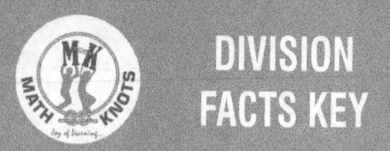
DP Exercise 4

1.	4 ÷ 4 =	1		1 × 4 = 4
2.	8 ÷ 4 =	2		2 × 4 = 8
3.	12 ÷ 4 =	3		3 × 4 = 12
4.	16 ÷ 4 =	4		4 × 4 = 16
5.	20 ÷ 4 =	5		5 × 4 = 20
6.	24 ÷ 4 =	6		6 × 4 = 24
7.	28 ÷ 4 =	7		7 × 4 = 28
8.	32 ÷ 4 =	8		8 × 4 = 32
9.	36 ÷ 4 =	9		9 × 4 = 36
10.	40 ÷ 4 =	10		10 × 4 = 40
11.	44 ÷ 4 =	11		11 × 4 = 44
12.	48 ÷ 4 =	12		12 × 4 = 48

Did you know division by 4 means
dividing the given number into 4 equal halfs ?

www.math-knots.com

DP Exercise 5

1.	5 ÷ 5 =	1	
2.	10 ÷ 5 =	2	
3.	15 ÷ 5 =	3	
4.	20 ÷ 5 =	4	
5.	25 ÷ 5 =	5	
6.	30 ÷ 5 =	6	
7.	35 ÷ 5 =	7	
8.	40 ÷ 5 =	8	
9.	45 ÷ 5 =	9	
10.	50 ÷ 5 =	10	
11.	55 ÷ 5 =	11	
12.	60 ÷ 5 =	12	

1	×	5	=	5
2	×	5	=	10
3	×	5	=	15
4	×	5	=	20
5	×	5	=	25
6	×	5	=	30
7	×	5	=	35
8	×	5	=	40
9	×	5	=	45
10	×	5	=	50
11	×	5	=	55
12	×	5	=	60

Did you know division by 5 means
dividing the given number into 5 equal halfs ?

www.math-knots.com

DP Exercise 6

1.	6 ÷ 6 =	1		1	× 6 =	6
2.	12 ÷ 6 =	2		2	× 6 =	12
3.	18 ÷ 6 =	3		3	× 6 =	18
4.	24 ÷ 6 =	4		4	× 6 =	24
5.	30 ÷ 6 =	5		5	× 6 =	30
6.	36 ÷ 6 =	6		6	× 6 =	36
7.	42 ÷ 6 =	7		7	× 6 =	42
8.	48 ÷ 6 =	8		8	× 6 =	48
9.	54 ÷ 6 =	9		9	× 6 =	54
10.	60 ÷ 6 =	10		10	× 6 =	60
11.	66 ÷ 6 =	11		11	× 6 =	66
12.	72 ÷ 6 =	12		12	× 6 =	72

Did you know division by 6 means
dividing the given number into 6 equal halfs ?

www.math-knots.com

DP Exercise 7

1.	$7 \div 7 =$	1
2.	$14 \div 7 =$	2
3.	$21 \div 7 =$	3
4.	$28 \div 7 =$	4
5.	$35 \div 7 =$	5
6.	$42 \div 7 =$	6
7.	$49 \div 7 =$	7
8.	$56 \div 7 =$	8
9.	$63 \div 7 =$	9
10.	$70 \div 7 =$	10
11.	$77 \div 7 =$	11
12.	$84 \div 7 =$	12

1	$\times$ 7 $=$	7
2	$\times$ 7 $=$	14
3	$\times$ 7 $=$	21
4	$\times$ 7 $=$	28
5	$\times$ 7 $=$	35
6	$\times$ 7 $=$	42
7	$\times$ 7 $=$	49
8	$\times$ 7 $=$	56
9	$\times$ 7 $=$	63
10	$\times$ 7 $=$	70
11	$\times$ 7 $=$	77
12	$\times$ 7 $=$	84

Did you know division by 7 means
dividing the given number into 7 equal halfs ?

www.math-knots.com

DP Exercise 8

1.	$8 \div 8 =$	1	
2.	$16 \div 8 =$	2	
3.	$24 \div 8 =$	3	
4.	$32 \div 8 =$	4	
5.	$40 \div 8 =$	5	
6.	$48 \div 8 =$	6	
7.	$56 \div 8 =$	7	
8.	$64 \div 8 =$	8	
9.	$72 \div 8 =$	9	
10.	$80 \div 8 =$	10	
11.	$88 \div 8 =$	11	
12.	$96 \div 8 =$	12	

1	$\times$	8	$= 8$
2	$\times$	8	$= 16$
3	$\times$	8	$= 24$
4	$\times$	8	$= 32$
5	$\times$	8	$= 40$
6	$\times$	8	$= 48$
7	$\times$	8	$= 56$
8	$\times$	8	$= 64$
9	$\times$	8	$= 72$
10	$\times$	8	$= 80$
11	$\times$	8	$= 88$
12	$\times$	8	$= 96$

Did you know division by 8 means dividing the given number into 8 equal halfs ?

www.math-knots.com

DP Exercise 9

1.	9 ÷ 9 =	1		1	×	9	=	9
2.	18 ÷ 9 =	2		2	×	9	=	18
3.	27 ÷ 9 =	3		3	×	9	=	27
4.	36 ÷ 9 =	4		4	×	9	=	36
5.	45 ÷ 9 =	5		5	×	9	=	45
6.	54 ÷ 9 =	6		6	×	9	=	54
7.	63 ÷ 9 =	7		7	×	9	=	63
8.	72 ÷ 9 =	8		8	×	9	=	72
9.	81 ÷ 9 =	9		9	×	9	=	81
10.	90 ÷ 9 =	10		10	×	9	=	90
11.	99 ÷ 9 =	11		11	×	9	=	99
12.	108 ÷ 9 =	12		12	×	9	=	108

Did you know division by 9 means
dividing the given number into 9 equal halfs ?

www.math-knots.com

DP Exercise 10

1.	10 ÷ 10 =	1	1 × 10 = 10
2.	20 ÷ 10 =	2	2 × 10 = 20
3.	30 ÷ 10 =	3	3 × 10 = 30
4.	40 ÷ 10 =	4	4 × 10 = 40
5.	50 ÷ 10 =	5	5 × 10 = 50
6.	60 ÷ 10 =	6	6 × 10 = 60
7.	70 ÷ 10 =	7	7 × 10 = 70
8.	80 ÷ 10 =	8	8 × 10 = 80
9.	90 ÷ 10 =	9	9 × 10 = 90
10.	100 ÷ 10 =	10	10 × 10 = 100
11.	110 ÷ 10 =	11	11 × 10 = 110
12.	120 ÷ 10 =	12	12 × 10 = 120

Did you know division by 10 means
dividing the given number into 10 equal halfs ?

www.math-knots.com

DP Exercise 11

1.	11 ÷ 11 =	1
2.	22 ÷ 11 =	2
3.	33 ÷ 11 =	3
4.	44 ÷ 11 =	4
5.	55 ÷ 11 =	5
6.	66 ÷ 11 =	6
7.	77 ÷ 11 =	7
8.	88 ÷ 11 =	8
9.	99 ÷ 11 =	9
10.	110 ÷ 11 =	10
11.	121 ÷ 11 =	11
12.	132 ÷ 11 =	12

1 ×	11	= 11
2 ×	11	= 22
3 ×	11	= 33
4 ×	11	= 44
5 ×	11	= 55
6 ×	11	= 66
7 ×	11	= 77
8 ×	11	= 88
9 ×	11	= 99
10 ×	11	= 110
11 ×	11	= 121
12 ×	11	= 132

Did you know division by 11 means
dividing the given number into 11 equal halfs ?

www.math-knots.com

DP Exercise 12

1.	12 ÷ 12 =	1
2.	24 ÷ 12 =	2
3.	36 ÷ 12 =	3
4.	48 ÷ 12 =	4
5.	60 ÷ 12 =	5
6.	72 ÷ 12 =	6
7.	84 ÷ 12 =	7
8.	96 ÷ 12 =	8
9.	108 ÷ 12 =	9
10.	120 ÷ 12 =	10
11.	132 ÷ 12 =	11
12.	144 ÷ 12 =	12

1	× 12 = 12
2	× 12 = 24
3	× 12 = 36
4	× 12 = 48
5	× 12 = 60
6	× 12 = 72
7	× 12 = 84
8	× 12 = 96
9	× 12 = 108
10	× 12 = 120
11	× 12 = 132
12	× 12 = 144

Did you know division by 12 means
dividing the given number into 12 equal halfs ?

www.math-knots.com

DP Exercise 13

1.	13 ÷ 13 =	1
2.	26 ÷ 13 =	2
3.	39 ÷ 13 =	3
4.	52 ÷ 13 =	4
5.	65 ÷ 13 =	5
6.	78 ÷ 13 =	6
7.	91 ÷ 13 =	7
8.	104 ÷ 13 =	8
9.	117 ÷ 13 =	9
10.	130 ÷ 13 =	10
11.	143 ÷ 13 =	11
12.	156 ÷ 13 =	12

1 × 13 = 13
2 × 13 = 26
3 × 13 = 39
4 × 13 = 52
5 × 13 = 65
6 × 13 = 78
7 × 13 = 91
8 × 13 = 104
9 × 13 = 117
10 × 13 = 130
11 × 13 = 143
12 × 13 = 156

Did You Know...?

Did you know division by 13 means dividing the given number into 13 equal halfs ?

www.math-knots.com

DP Exercise 14

1.	14 ÷ 14 =	1
2.	28 ÷ 14 =	2
3.	42 ÷ 14 =	3
4.	56 ÷ 14 =	4
5.	70 ÷ 14 =	5
6.	84 ÷ 14 =	6
7.	98 ÷ 14 =	7
8.	112 ÷ 14 =	8
9.	126 ÷ 14 =	9
10.	140 ÷ 14 =	10
11.	154 ÷ 14 =	11
12.	168 ÷ 14 =	12

1	× 14	= 14
2	× 14	= 28
3	× 14	= 42
4	× 14	= 56
5	× 14	= 70
6	× 14	= 84
7	× 14	= 98
8	× 14	= 112
9	× 14	= 126
10	× 14	= 140
11	× 14	= 154
12	× 14	= 168

Did you know division by 14 means
dividing the given number into 14 equal halfs ?

www.math-knots.com

DIVISION FACTS KEY

Practice

DP Exercise 15

#	Division	Answer		#	Multiplication	Answer
1.	15 ÷ 15 =	1		1 × 15	= 15
2.	30 ÷ 15 =	2		2 × 15	= 30
3.	45 ÷ 15 =	3		3 × 15	= 45
4.	60 ÷ 15 =	4		4 × 15	= 60
5.	75 ÷ 15 =	5		5 × 15	= 75
6.	90 ÷ 15 =	6		6 × 15	= 90
7.	105 ÷ 15 =	7		7 × 15	= 105
8.	120 ÷ 15 =	8		8 × 15	= 120
9.	135 ÷ 15 =	9		9 × 15	= 135
10.	150 ÷ 15 =	10		10 × 15	= 150
11.	165 ÷ 15 =	11		11 × 15	= 165
12.	180 ÷ 15 =	12		12 × 15	= 180

Did you know division by 15 means
dividing the given number into 15 equal halfs ?

www.math-knots.com

DP Exercise 16

1.	16 ÷ 16	=	1
2.	32 ÷ 16	=	2
3.	48 ÷ 16	=	3
4.	64 ÷ 16	=	4
5.	80 ÷ 16	=	5
6.	96 ÷ 16	=	6
7.	112 ÷ 16	=	7
8.	128 ÷ 16	=	8
9.	144 ÷ 16	=	9
10.	160 ÷ 16	=	10
11.	176 ÷ 16	=	11
12.	192 ÷ 16	=	12

1	× 16	= 16
2	× 16	= 32
3	× 16	= 48
4	× 16	= 64
5	× 16	= 80
6	× 16	= 96
7	× 16	= 112
8	× 16	= 128
9	× 16	= 144
10	× 16	= 160
11	× 16	= 176
12	× 16	= 192

Did you know division by 16 means
dividing the given number into 16 equal halfs ?

www.math-knots.com

DP Exercise 17

1.	17 ÷ 17 =	1	
2.	34 ÷ 17 =	2	
3.	51 ÷ 17 =	3	
4.	68 ÷ 17 =	4	
5.	85 ÷ 17 =	5	
6.	102 ÷ 17 =	6	
7.	119 ÷ 17 =	7	
8.	136 ÷ 17 =	8	
9.	153 ÷ 17 =	9	
10.	170 ÷ 17 =	10	
11.	187 ÷ 17 =	11	
12.	204 ÷ 17 =	12	

1 × 17 = 17
2 × 17 = 34
3 × 17 = 51
4 × 17 = 68
5 × 17 = 85
6 × 17 = 102
7 × 17 = 119
8 × 17 = 136
9 × 17 = 153
10 × 17 = 170
11 × 17 = 187
12 × 17 = 204

Did you know division by 17 means
dividing the given number into 17 equal halfs ?

www.math-knots.com

DP Exercise 18

1.	18 ÷ 18 =	1
2.	36 ÷ 18 =	2
3.	54 ÷ 18 =	3
4.	72 ÷ 18 =	4
5.	90 ÷ 18 =	5
6.	108 ÷ 18 =	6
7.	126 ÷ 18 =	7
8.	144 ÷ 18 =	8
9.	162 ÷ 18 =	9
10.	180 ÷ 18 =	10
11.	198 ÷ 18 =	11
12.	216 ÷ 18 =	12

1 × 18 = 18
2 × 18 = 36
3 × 18 = 54
4 × 18 = 72
5 × 18 = 90
6 × 18 = 108
7 × 18 = 126
8 × 18 = 142
9 × 18 = 162
10 × 18 = 180
11 × 18 = 198
12 × 18 = 216

Did you know division by 18 means dividing the given number into 18 equal halfs ?

www.math-knots.com

DP Exercise 19

1.	$19 \div 19 =$	1
2.	$38 \div 19 =$	2
3.	$57 \div 19 =$	3
4.	$76 \div 19 =$	4
5.	$95 \div 19 =$	5
6.	$114 \div 19 =$	6
7.	$133 \div 19 =$	7
8.	$152 \div 19 =$	8
9.	$171 \div 19 =$	9
10.	$190 \div 19 =$	10
11.	$209 \div 19 =$	11
12.	$228 \div 19 =$	12

1	$\times$	19	$= 19$
2	$\times$	19	$= 38$
3	$\times$	19	$= 57$
4	$\times$	19	$= 76$
5	$\times$	19	$= 95$
6	$\times$	19	$= 114$
7	$\times$	19	$= 133$
8	$\times$	19	$= 152$
9	$\times$	19	$= 171$
10	$\times$	19	$= 190$
11	$\times$	19	$= 209$
12	$\times$	19	$= 228$

Did you know division by 19 means
dividing the given number into 19 equal halfs ?

www.math-knots.com

DP Exercise 20

1.	20 ÷ 20 =	1
2.	40 ÷ 20 =	2
3.	60 ÷ 20 =	3
4.	80 ÷ 20 =	4
5.	100 ÷ 20 =	5
6.	120 ÷ 20 =	6
7.	140 ÷ 20 =	7
8.	160 ÷ 20 =	8
9.	180 ÷ 20 =	9
10.	200 ÷ 20 =	10
11.	220 ÷ 20 =	11
12.	240 ÷ 20 =	12

1	×	20	= 20
2	×	20	= 40
3	×	20	= 60
4	×	20	= 80
5	×	20	= 100
6	×	20	= 120
7	×	20	= 140
8	×	20	= 160
9	×	20	= 180
10	×	20	= 200
11	×	20	= 220
12	×	20	= 240

Did you know division by 20 means
dividing the given number into 20 equal halfs ?

www.math-knots.com

DP Exercise 21

1.	21 ÷ 21 =	1
2.	42 ÷ 21 =	2
3.	63 ÷ 21 =	3
4.	84 ÷ 21 =	4
5.	105 ÷ 21 =	5
6.	126 ÷ 21 =	6
7.	147 ÷ 21 =	7
8.	168 ÷ 21 =	8
9.	189 ÷ 21 =	9
10.	210 ÷ 21 =	10
11.	231 ÷ 21 =	11
12.	252 ÷ 21 =	12

1	×	21	= 21
2	×	21	= 42
3	×	21	= 63
4	×	21	= 84
5	×	21	= 105
6	×	21	= 126
7	×	21	= 147
8	×	21	= 168
9	×	21	= 189
10	×	21	= 210
11	×	21	= 231
12	×	21	= 252

Did you know division by 21 means
dividing the given number into 21 equal halfs ?

www.math-knots.com

DP Exercise 22

1.	22 ÷ 22 =	1
2.	44 ÷ 22 =	2
3.	66 ÷ 22 =	3
4.	88 ÷ 22 =	4
5.	110 ÷ 22 =	5
6.	132 ÷ 22 =	6
7.	154 ÷ 22 =	7
8.	176 ÷ 22 =	8
9.	198 ÷ 22 =	9
10.	220 ÷ 22 =	10
11.	242 ÷ 22 =	11
12.	264 ÷ 22 =	12

1	× 22	= 22
2	× 22	= 44
3	× 22	= 66
4	× 22	= 88
5	× 22	= 110
6	× 22	= 132
7	× 22	= 154
8	× 22	= 176
9	× 22	= 198
10	× 22	= 220
11	× 22	= 242
12	× 22	= 264

Did you know division by 22 means dividing the given number into 22 equal halfs ?

www.math-knots.com

DP Exercise 23

1.	23 ÷ 23 =	1
2.	46 ÷ 23 =	2
3.	69 ÷ 23 =	3
4.	92 ÷ 22 =	4
5.	115 ÷ 23 =	5
6.	138 ÷ 23 =	6
7.	161 ÷ 23 =	7
8.	184 ÷ 23 =	8
9.	207 ÷ 23 =	9
10.	230 ÷ 23 =	10
11.	253 ÷ 23 =	11
12.	276 ÷ 23 =	12

1	× 23	= 23
2	× 23	= 46
3	× 23	= 69
4	× 23	= 92
5	× 23	= 115
6	× 23	= 138
7	× 23	= 161
8	× 23	= 184
9	× 23	= 207
10	× 23	= 230
11	× 23	= 253
12	× 23	= 276

Did you know division by 23 means
dividing the given number into 23 equal halfs ?

www.math-knots.com

DP Exercise 24

1.	$24 \div 24 =$	1
2.	$48 \div 24 =$	2
3.	$72 \div 24 =$	3
4.	$96 \div 24 =$	4
5.	$120 \div 24 =$	5
6.	$144 \div 24 =$	6
7.	$168 \div 24 =$	7
8.	$192 \div 24 =$	8
9.	$216 \div 24 =$	9
10.	$240 \div 24 =$	10
11.	$264 \div 24 =$	11
12.	$288 \div 24 =$	12

1	$\times$	24	$= 24$
2	$\times$	24	$= 48$
3	$\times$	24	$= 72$
4	$\times$	24	$= 96$
5	$\times$	24	$= 120$
6	$\times$	24	$= 144$
7	$\times$	24	$= 168$
8	$\times$	24	$= 192$
9	$\times$	24	$= 216$
10	$\times$	24	$= 240$
11	$\times$	24	$= 264$
12	$\times$	24	$= 288$

Did you know division by 24 means
dividing the given number into 24 equal halfs ?

www.math-knots.com

DP Exercise 25

1.	25 ÷ 25 =	1
2.	50 ÷ 25 =	2
3.	75 ÷ 25 =	3
4.	100 ÷ 25 =	4
5.	125 ÷ 25 =	5
6.	150 ÷ 25 =	6
7.	175 ÷ 25 =	7
8.	200 ÷ 25 =	8
9.	225 ÷ 25 =	9
10.	250 ÷ 25 =	10
11.	275 ÷ 25 =	11
12.	300 ÷ 25 =	12

1 × 25	= 25
2 × 25	= 50
3 × 25	= 75
4 × 25	= 100
5 × 25	= 125
6 × 25	= 150
7 × 25	= 175
8 × 25	= 200
9 × 25	= 225
10 × 25	= 250
11 × 25	= 275
12 × 25	= 300

Did you know division by 25 means
dividing the given number into 25 equal halfs ?

www.math-knots.com

DP Exercise 26

1.	67	÷ 1	=	67
2.	44	÷ 2	=	22
3.	99	÷ 3	=	33
4.	80	÷ 4	=	20
5.	95	÷ 5	=	19
6.	126	÷ 6	=	21
7.	49	÷ 7	=	7
8.	96	÷ 8	=	12
9.	189	÷ 9	=	21
10.	800	÷ 10	=	80
11.	1331	÷ 11	=	121
12.	1728	÷ 12	=	144

1	×	67	=	67
2	×	22	=	44
3	×	33	=	99
4	×	20	=	80
5	×	19	=	95
6	×	21	=	126
7	×	7	=	49
8	×	12	=	96
9	×	21	=	189
10	×	80	=	800
11	×	121	=	1331
12	×	144	=	1728

Did you know operations , division and multiplication are opposites of each other ?

www.math-knots.com

DP Exercise 27

1.	591	÷ 1	=	591
2.	404	÷ 2	=	202
3.	303	÷ 3	=	101
4.	84	÷ 4	=	21
5.	125	÷ 5	=	25
6.	246	÷ 6	=	41
7.	777	÷ 7	=	111
8.	160	÷ 8	=	20
9.	279	÷ 9	=	31
10.	2010	÷ 10	=	201
11.	11011	÷ 11	=	1001
12.	121212	÷ 12	=	10101

1	×	591	=	591
2	×	202	=	404
3	×	101	=	303
4	×	21	=	84
5	×	25	=	125
6	×	41	=	246
7	×	111	=	777
8	×	20	=	160
9	×	31	=	279
10	×	201	=	2010
11	×	1001	=	11011
12	×	10101	=	121212

Did you know operations , division and multiplication are opposites of each other ?

www.math-knots.com

DP Exercise 28

(A) $1\overline{)2}$

Ans : $1\overline{)\overset{2}{2}}$

(F) $1\overline{)7}$

Ans : $1\overline{)\overset{7}{7}}$

(K) $1\overline{)12}$

Ans : $1\overline{)\overset{12}{12}}$

(B) $1\overline{)3}$

Ans : $1\overline{)\overset{3}{3}}$

(G) $1\overline{)8}$

Ans : $1\overline{)\overset{8}{8}}$

(L) $1\overline{)13}$

Ans : $1\overline{)\overset{13}{13}}$

(C) $1\overline{)4}$

Ans : $1\overline{)\overset{4}{4}}$

(H) $1\overline{)9}$

Ans : $1\overline{)\overset{9}{9}}$

(M) $1\overline{)14}$

Ans : $1\overline{)\overset{14}{14}}$

(D) $1\overline{)5}$

Ans : $1\overline{)\overset{5}{5}}$

(I) $1\overline{)10}$

Ans : $1\overline{)\overset{10}{10}}$

(N) $1\overline{)15}$

Ans : $1\overline{)\overset{15}{15}}$

(E) $1\overline{)6}$

Ans : $1\overline{)\overset{6}{6}}$

(J) $1\overline{)11}$

Ans : $1\overline{)\overset{11}{11}}$

(O) $1\overline{)16}$

Ans : $1\overline{)\overset{16}{16}}$

www.math-knots.com

DP Exercise 29

(A) $2\overline{)2}$

Ans : $2\overline{)\overset{1}{2}}$

(B) $2\overline{)4}$

Ans : $2\overline{)\overset{2}{4}}$

(C) $2\overline{)6}$

Ans : $2\overline{)\overset{3}{6}}$

(D) $2\overline{)8}$

Ans : $2\overline{)\overset{4}{8}}$

(E) $2\overline{)10}$

Ans : $2\overline{)\overset{5}{10}}$

(F) $2\overline{)12}$

Ans : $2\overline{)\overset{6}{12}}$

(G) $2\overline{)14}$

Ans : $2\overline{)\overset{7}{14}}$

(H) $2\overline{)16}$

Ans : $2\overline{)\overset{8}{16}}$

(I) $2\overline{)18}$

Ans : $2\overline{)\overset{9}{18}}$

(J) $2\overline{)20}$

Ans : $2\overline{)\overset{10}{20}}$

(K) $2\overline{)22}$

Ans : $2\overline{)\overset{11}{22}}$

(L) $2\overline{)24}$

Ans : $2\overline{)\overset{12}{24}}$

(M) $2\overline{)26}$

Ans : $2\overline{)\overset{13}{26}}$

(N) $2\overline{)28}$

Ans : $2\overline{)\overset{14}{28}}$

(O) $2\overline{)30}$

Ans : $2\overline{)\overset{15}{30}}$

DP Exercise 30

(A) $3\overline{)3}$

Ans : $3\overline{)\,\overset{1}{3}}$

(F) $3\overline{)18}$

Ans : $3\overline{)\,\overset{6}{18}}$

(K) $3\overline{)33}$

Ans : $3\overline{)\,\overset{11}{33}}$

(B) $3\overline{)6}$

Ans : $3\overline{)\,\overset{2}{6}}$

(G) $3\overline{)21}$

Ans : $3\overline{)\,\overset{7}{21}}$

(L) $3\overline{)36}$

Ans : $3\overline{)\,\overset{12}{36}}$

(C) $3\overline{)9}$

Ans : $3\overline{)\,\overset{3}{9}}$

(H) $3\overline{)24}$

Ans : $3\overline{)\,\overset{8}{24}}$

(M) $3\overline{)39}$

Ans : $3\overline{)\,\overset{13}{39}}$

(D) $3\overline{)12}$

Ans : $3\overline{)\,\overset{4}{12}}$

(I) $3\overline{)27}$

Ans : $3\overline{)\,\overset{9}{27}}$

(N) $3\overline{)42}$

Ans : $3\overline{)\,\overset{14}{42}}$

(E) $3\overline{)15}$

Ans : $3\overline{)\,\overset{5}{15}}$

(J) $3\overline{)30}$

Ans : $3\overline{)\,\overset{10}{30}}$

(O) $3\overline{)45}$

Ans : $3\overline{)\,\overset{15}{45}}$

DP Exercise 31

(A) 4⟌4

Ans : 4⟌4̄ with 1 above

(B) 4⟌8

Ans : 4⟌8̄ with 2 above

(C) 4⟌12

Ans : 4⟌12̄ with 3 above

(D) 4⟌16

Ans : 4⟌16̄ with 4 above

(E) 4⟌20

Ans : 4⟌20̄ with 5 above

(F) 4⟌24

Ans : 4⟌24̄ with 6 above

(G) 4⟌28

Ans : 4⟌28̄ with 7 above

(H) 4⟌32

Ans : 4⟌32̄ with 8 above

(I) 4⟌36

Ans : 4⟌36̄ with 9 above

(J) 4⟌40

Ans : 4⟌40̄ with 10 above

(K) 4⟌44

Ans : 4⟌44̄ with 11 above

(L) 4⟌48

Ans : 4⟌48̄ with 12 above

(M) 4⟌52

Ans : 4⟌52̄ with 13 above

(N) 4⟌56

Ans : 4⟌56̄ with 14 above

(O) 4⟌60

Ans : 4⟌60̄ with 15 above

DP Exercise 32

(A) $5 \overline{) 5}$

Ans : $5 \overline{) 5}^{\,1}$

(B) $5 \overline{) 10}$

Ans : $5 \overline{) 10}^{\,2}$

(C) $5 \overline{) 15}$

Ans : $5 \overline{) 15}^{\,3}$

(D) $5 \overline{) 20}$

Ans : $5 \overline{) 20}^{\,4}$

(E) $5 \overline{) 25}$

Ans : $5 \overline{) 25}^{\,5}$

(F) $5 \overline{) 30}$

Ans : $5 \overline{) 30}^{\,6}$

(G) $5 \overline{) 35}$

Ans : $5 \overline{) 35}^{\,7}$

(H) $5 \overline{) 40}$

Ans : $5 \overline{) 40}^{\,8}$

(I) $5 \overline{) 45}$

Ans : $5 \overline{) 45}^{\,9}$

(J) $5 \overline{) 50}$

Ans : $5 \overline{) 50}^{\,10}$

(K) $5 \overline{) 55}$

Ans : $5 \overline{) 55}^{\,11}$

(L) $5 \overline{) 60}$

Ans : $5 \overline{) 60}^{\,12}$

(M) $5 \overline{) 65}$

Ans : $5 \overline{) 65}^{\,13}$

(N) $5 \overline{) 70}$

Ans : $5 \overline{) 70}^{\,14}$

(O) $5 \overline{) 75}$

Ans : $5 \overline{) 75}^{\,15}$

www.math-knots.com

DP Exercise 33

(A) $6\overline{)6}$

Ans : $6\overline{)6}^{\,1}$

(B) $6\overline{)12}$

Ans : $6\overline{)12}^{\,2}$

(C) $6\overline{)18}$

Ans : $6\overline{)18}^{\,3}$

(D) $6\overline{)24}$

Ans : $6\overline{)24}^{\,4}$

(E) $6\overline{)30}$

Ans : $6\overline{)30}^{\,5}$

(F) $6\overline{)36}$

Ans : $6\overline{)36}^{\,6}$

(G) $6\overline{)42}$

Ans : $6\overline{)42}^{\,7}$

(H) $6\overline{)48}$

Ans : $6\overline{)48}^{\,8}$

(I) $6\overline{)54}$

Ans : $6\overline{)54}^{\,9}$

(J) $6\overline{)60}$

Ans : $6\overline{)60}^{\,10}$

(K) $6\overline{)66}$

Ans : $6\overline{)66}^{\,11}$

(L) $6\overline{)72}$

Ans : $6\overline{)72}^{\,12}$

(M) $6\overline{)78}$

Ans : $6\overline{)78}^{\,13}$

(N) $6\overline{)84}$

Ans : $6\overline{)84}^{\,14}$

(O) $6\overline{)90}$

Ans : $6\overline{)90}^{\,15}$

www.math-knots.com

DP Exercise 34

(A) 7⟌7

Ans : 7⟌7̄ with 1 on top

(F) 7⟌42

Ans : 7⟌42 with 6 on top

(K) 7⟌77

Ans : 7⟌77 with 11 on top

(B) 7⟌14

Ans : 7⟌14 with 2 on top

(G) 7⟌49

Ans : 7⟌49 with 7 on top

(L) 7⟌84

Ans : 7⟌84 with 12 on top

(C) 7⟌21

Ans : 7⟌21 with 3 on top

(H) 7⟌56

Ans : 7⟌56 with 8 on top

(M) 7⟌91

Ans : 7⟌91 with 13 on top

(D) 7⟌28

Ans : 7⟌28 with 4 on top

(I) 7⟌63

Ans : 7⟌63 with 9 on top

(N) 7⟌98

Ans : 7⟌98 with 14 on top

(E) 7⟌35

Ans : 7⟌35 with 5 on top

(J) 7⟌70

Ans : 7⟌70 with 10 on top

(O) 7⟌105

Ans : 7⟌105 with 15 on top

DP Exercise 35

(A) $8\overline{)8}$

Ans : $8\overline{)8}^{\,1}$

(B) $8\overline{)16}$

Ans : $8\overline{)16}^{\,2}$

(C) $8\overline{)24}$

Ans : $8\overline{)24}^{\,3}$

(D) $8\overline{)32}$

Ans : $8\overline{)32}^{\,4}$

(E) $8\overline{)40}$

Ans : $8\overline{)40}^{\,5}$

(F) $8\overline{)48}$

Ans : $8\overline{)48}^{\,6}$

(G) $8\overline{)56}$

Ans : $8\overline{)56}^{\,7}$

(H) $8\overline{)64}$

Ans : $8\overline{)64}^{\,8}$

(I) $8\overline{)72}$

Ans : $8\overline{)72}^{\,9}$

(J) $8\overline{)80}$

Ans : $8\overline{)80}^{\,10}$

(K) $8\overline{)88}$

Ans : $8\overline{)88}^{\,11}$

(L) $8\overline{)96}$

Ans : $8\overline{)96}^{\,12}$

(M) $8\overline{)104}$

Ans : $8\overline{)104}^{\,13}$

(N) $8\overline{)112}$

Ans : $8\overline{)112}^{\,14}$

(O) $8\overline{)120}$

Ans : $8\overline{)120}^{\,15}$

www.math-knots.com

DP Exercise 36

(A) $9\overline{)9}$

Ans : $9\overset{1}{\overline{)9}}$

(F) $9\overline{)54}$

Ans : $9\overset{6}{\overline{)54}}$

(K) $9\overline{)99}$

Ans : $9\overset{11}{\overline{)99}}$

(B) $9\overline{)18}$

Ans : $9\overset{2}{\overline{)18}}$

(G) $9\overline{)63}$

Ans : $9\overset{7}{\overline{)63}}$

(L) $9\overline{)108}$

Ans : $9\overset{12}{\overline{)108}}$

(C) $9\overline{)27}$

Ans : $9\overset{3}{\overline{)27}}$

(H) $9\overline{)72}$

Ans : $9\overset{8}{\overline{)72}}$

(M) $9\overline{)117}$

Ans : $9\overset{13}{\overline{)117}}$

(D) $9\overline{)36}$

Ans : $9\overset{4}{\overline{)36}}$

(I) $9\overline{)81}$

Ans : $9\overset{9}{\overline{)81}}$

(N) $9\overline{)126}$

Ans : $9\overset{14}{\overline{)126}}$

(E) $9\overline{)45}$

Ans : $9\overset{5}{\overline{)45}}$

(J) $9\overline{)90}$

Ans : $9\overset{10}{\overline{)90}}$

(O) $9\overline{)135}$

Ans : $9\overset{15}{\overline{)135}}$

www.math-knots.com

DP Exercise 37

(A) $10\overline{)10}$

Ans : $10\overline{)10}^{\,1}$

(B) $10\overline{)20}$

Ans : $10\overline{)20}^{\,2}$

(C) $10\overline{)30}$

Ans : $10\overline{)30}^{\,3}$

(D) $10\overline{)40}$

Ans : $10\overline{)40}^{\,4}$

(E) $10\overline{)50}$

Ans : $10\overline{)50}^{\,5}$

(F) $10\overline{)60}$

Ans : $10\overline{)60}^{\,6}$

(G) $10\overline{)70}$

Ans : $10\overline{)70}^{\,7}$

(H) $10\overline{)80}$

Ans : $10\overline{)80}^{\,8}$

(I) $10\overline{)90}$

Ans : $10\overline{)90}^{\,9}$

(J) $10\overline{)100}$

Ans : $10\overline{)100}^{\,10}$

(K) $10\overline{)110}$

Ans : $10\overline{)110}^{\,11}$

(L) $10\overline{)120}$

Ans : $10\overline{)120}^{\,12}$

(M) $10\overline{)130}$

Ans : $10\overline{)130}^{\,13}$

(N) $10\overline{)140}$

Ans : $10\overline{)140}^{\,14}$

(O) $10\overline{)150}$

Ans : $10\overline{)150}^{\,15}$

DP Exercise 38

(A) 11)‾1‾1‾

Ans : 11)‾1‾1‾ → 1

(F) 11)‾6‾6‾

Ans : 11)‾6‾6‾ → 6

(K) 11)‾1‾2‾1‾

Ans : 11)‾1‾2‾1‾ → 11

(B) 11)‾2‾2‾

Ans : 11)‾2‾2‾ → 2

(G) 11)‾7‾7‾

Ans : 11)‾7‾7‾ → 7

(L) 11)‾1‾3‾2‾

Ans : 11)‾1‾3‾2‾ → 12

(C) 11)‾3‾3‾

Ans : 11)‾3‾3‾ → 3

(H) 11)‾8‾8‾

Ans : 11)‾8‾8‾ → 8

(M) 11)‾1‾4‾3‾

Ans : 11)‾1‾4‾3‾ → 13

(D) 11)‾4‾4‾

Ans : 11)‾4‾4‾ → 4

(I) 11)‾9‾9‾

Ans : 11)‾9‾9‾ → 9

(N) 11)‾1‾5‾4‾

Ans : 11)‾1‾5‾4‾ → 14

(E) 11)‾5‾5‾

Ans : 11)‾5‾5‾ → 5

(J) 11)‾1‾1‾0‾

Ans : 11)‾1‾1‾0‾ → 10

(O) 11)‾1‾6‾5‾

Ans : 11)‾1‾6‾5‾ → 15

DP Exercise 39

(A) $12\overline{)12}$

Ans : $12\overline{)\overset{1}{12}}$

(F) $12\overline{)72}$

Ans : $12\overline{)\overset{6}{72}}$

(K) $12\overline{)132}$

Ans : $12\overline{)\overset{11}{132}}$

(B) $12\overline{)24}$

Ans : $12\overline{)\overset{2}{24}}$

(G) $12\overline{)84}$

Ans : $12\overline{)\overset{7}{84}}$

(L) $12\overline{)144}$

Ans : $12\overline{)\overset{12}{144}}$

(C) $12\overline{)36}$

Ans : $12\overline{)\overset{3}{36}}$

(H) $12\overline{)96}$

Ans : $12\overline{)\overset{8}{96}}$

(M) $12\overline{)156}$

Ans : $12\overline{)\overset{13}{156}}$

(D) $12\overline{)48}$

Ans : $12\overline{)\overset{4}{48}}$

(I) $12\overline{)108}$

Ans : $12\overline{)\overset{9}{108}}$

(N) $12\overline{)168}$

Ans : $12\overline{)\overset{14}{168}}$

(E) $12\overline{)60}$

Ans : $12\overline{)\overset{5}{60}}$

(J) $12\overline{)120}$

Ans : $12\overline{)\overset{10}{120}}$

(O) $12\overline{)180}$

Ans : $12\overline{)\overset{15}{180}}$

www.math-knots.com

DP Exercise 40

(A) 13 ⟌ 13

Ans : 13 ⟌ 13 (1)

(B) 13 ⟌ 26

Ans : 13 ⟌ 26 (2)

(C) 13 ⟌ 39

Ans : 13 ⟌ 39 (3)

(D) 13 ⟌ 52

Ans : 13 ⟌ 52 (4)

(E) 13 ⟌ 65

Ans : 13 ⟌ 65 (5)

(F) 13 ⟌ 78

Ans : 13 ⟌ 78 (6)

(G) 13 ⟌ 91

Ans : 13 ⟌ 91 (7)

(H) 13 ⟌ 104

Ans : 13 ⟌ 104 (8)

(I) 13 ⟌ 117

Ans : 13 ⟌ 117 (9)

(J) 13 ⟌ 130

Ans : 13 ⟌ 130 (10)

(K) 13 ⟌ 143

Ans : 13 ⟌ 143 (11)

(L) 13 ⟌ 156

Ans : 13 ⟌ 156 (12)

(M) 13 ⟌ 169

Ans : 13 ⟌ 169 (13)

(N) 13 ⟌ 182

Ans : 13 ⟌ 182 (14)

(O) 13 ⟌ 195

Ans : 13 ⟌ 195 (15)

www.math-knots.com

DP Exercise 41

(A) $14\overline{)14}$

Ans : $14\overline{)\overset{1}{14}}$

(B) $14\overline{)28}$

Ans : $14\overline{)\overset{2}{28}}$

(C) $14\overline{)42}$

Ans : $14\overline{)\overset{3}{42}}$

(D) $14\overline{)56}$

Ans : $14\overline{)\overset{4}{56}}$

(E) $14\overline{)70}$

Ans : $14\overline{)\overset{5}{70}}$

(F) $14\overline{)84}$

Ans : $14\overline{)\overset{6}{84}}$

(G) $14\overline{)98}$

Ans : $14\overline{)\overset{7}{98}}$

(H) $14\overline{)112}$

Ans : $14\overline{)\overset{8}{112}}$

(I) $14\overline{)126}$

Ans : $14\overline{)\overset{9}{126}}$

(J) $14\overline{)140}$

Ans : $14\overline{)\overset{10}{140}}$

(K) $14\overline{)154}$

Ans : $14\overline{)\overset{11}{154}}$

(L) $14\overline{)168}$

Ans : $14\overline{)\overset{12}{168}}$

(M) $14\overline{)182}$

Ans : $14\overline{)\overset{13}{182}}$

(N) $14\overline{)196}$

Ans : $14\overline{)\overset{14}{196}}$

(O) $14\overline{)210}$

Ans : $14\overline{)\overset{15}{210}}$

DP Exercise 42

(A) 15⟌15

Ans : 15⟌15 → 1

(B) 15⟌30

Ans : 15⟌30 → 2

(C) 15⟌45

Ans : 15⟌45 → 3

(D) 15⟌60

Ans : 15⟌60 → 4

(E) 15⟌75

Ans : 15⟌75 → 5

(F) 15⟌90

Ans : 15⟌90 → 6

(G) 15⟌105

Ans : 15⟌105 → 7

(H) 15⟌120

Ans : 15⟌120 → 8

(I) 15⟌135

Ans : 15⟌135 → 9

(J) 15⟌150

Ans : 15⟌150 → 10

(K) 15⟌165

Ans : 15⟌165 → 11

(L) 15⟌180

Ans : 15⟌180 → 12

(M) 15⟌195

Ans : 15⟌195 → 13

(N) 15⟌210

Ans : 15⟌210 → 14

(O) 15⟌225

Ans : 15⟌225 → 15

www.math-knots.com

DP Exercise 43

(A) 16 ⟌ 16

Ans : 16 ⟌ 16 → 1

(F) 16 ⟌ 96

Ans : 16 ⟌ 96 → 6

(K) 16 ⟌ 176

Ans : 16 ⟌ 176 → 11

(B) 16 ⟌ 32

Ans : 16 ⟌ 32 → 2

(G) 16 ⟌ 112

Ans : 16 ⟌ 112 → 7

(L) 16 ⟌ 192

Ans : 16 ⟌ 192 → 12

(C) 16 ⟌ 48

Ans : 16 ⟌ 48 → 3

(H) 16 ⟌ 128

Ans : 16 ⟌ 128 → 8

(M) 16 ⟌ 208

Ans : 16 ⟌ 208 → 13

(D) 16 ⟌ 64

Ans : 16 ⟌ 64 → 4

(I) 16 ⟌ 144

Ans : 16 ⟌ 144 → 9

(N) 16 ⟌ 224

Ans : 16 ⟌ 224 → 14

(E) 16 ⟌ 80

Ans : 16 ⟌ 80 → 5

(J) 16 ⟌ 160

Ans : 16 ⟌ 160 → 10

(O) 16 ⟌ 240

Ans : 16 ⟌ 240 → 15

www.math-knots.com

DP Exercise 44

(A)   17 ⟌ 17

Ans : 17 ⟌ 17 , quotient 1

(F)   17 ⟌ 102

Ans : 17 ⟌ 102 , quotient 6

(K)   17 ⟌ 187

Ans : 17 ⟌ 187 , quotient 11

(B)   17 ⟌ 34

Ans : 17 ⟌ 34 , quotient 2

(G)   17 ⟌ 119

Ans : 17 ⟌ 119 , quotient 7

(L)   17 ⟌ 204

Ans : 17 ⟌ 204 , quotient 12

(C)   17 ⟌ 51

Ans : 17 ⟌ 51 , quotient 3

(H)   17 ⟌ 136

Ans : 17 ⟌ 136 , quotient 8

(M)   17 ⟌ 221

Ans : 17 ⟌ 221 , quotient 13

(D)   17 ⟌ 68

Ans : 17 ⟌ 68 , quotient 4

(I)   17 ⟌ 153

Ans : 17 ⟌ 153 , quotient 9

(N)   17 ⟌ 238

Ans : 17 ⟌ 238 , quotient 14

(E)   17 ⟌ 85

Ans : 17 ⟌ 85 , quotient 5

(J)   17 ⟌ 170

Ans : 17 ⟌ 170 , quotient 10

(O)   17 ⟌ 255

Ans : 17 ⟌ 255 , quotient 15

www.math-knots.com

DP Exercise 45

(A)   18⟌18

Ans : 18⟌18 ⟍ 1

(F)   18⟌108

Ans : 18⟌108 ⟍ 6

(K)   18⟌198

Ans : 18⟌198 ⟍ 11

(B)   18⟌36

Ans : 18⟌36 ⟍ 2

(G)   18⟌126

Ans : 18⟌126 ⟍ 7

(L)   18⟌216

Ans : 18⟌216 ⟍ 12

(C)   18⟌54

Ans : 18⟌54 ⟍ 3

(H)   18⟌144

Ans : 18⟌144 ⟍ 8

(M)   18⟌234

Ans : 18⟌234 ⟍ 13

(D)   18⟌72

Ans : 18⟌72 ⟍ 4

(I)   18⟌162

Ans : 18⟌162 ⟍ 9

(N)   18⟌252

Ans : 18⟌252 ⟍ 14

(E)   18⟌90

Ans : 18⟌90 ⟍ 5

(J)   18⟌180

Ans : 18⟌180 ⟍ 10

(O)   18⟌270

Ans : 18⟌270 ⟍ 15

DP Exercise 46

(A) $19\overline{)19}$

Ans : $19\overline{)19}^{\,1}$

(B) $19\overline{)38}$

Ans : $19\overline{)38}^{\,2}$

(C) $19\overline{)57}$

Ans : $19\overline{)57}^{\,3}$

(D) $19\overline{)76}$

Ans : $19\overline{)76}^{\,4}$

(E) $19\overline{)95}$

Ans : $19\overline{)95}^{\,5}$

(F) $19\overline{)114}$

Ans : $19\overline{)114}^{\,6}$

(G) $19\overline{)133}$

Ans : $19\overline{)133}^{\,7}$

(H) $19\overline{)152}$

Ans : $19\overline{)152}^{\,8}$

(I) $19\overline{)171}$

Ans : $19\overline{)171}^{\,9}$

(J) $19\overline{)190}$

Ans : $19\overline{)190}^{\,10}$

(K) $19\overline{)209}$

Ans : $19\overline{)209}^{\,11}$

(L) $19\overline{)228}$

Ans : $19\overline{)228}^{\,12}$

(M) $19\overline{)247}$

Ans : $19\overline{)247}^{\,13}$

(N) $19\overline{)266}$

Ans : $19\overline{)266}^{\,14}$

(O) $19\overline{)285}$

Ans : $19\overline{)285}^{\,15}$

www.math-knots.com

DP Exercise 47

(A)   20⟌20

Ans : 20⟌20̄ ¹

(B)   20⟌40

Ans : 20⟌40̄ ²

(C)   20⟌60

Ans : 20⟌60̄ ³

(D)   20⟌80

Ans : 20⟌80̄ ⁴

(E)   20⟌100

Ans : 20⟌100̄ ⁵

(F)   20⟌120

Ans : 20⟌120̄ ⁶

(G)   20⟌140

Ans : 20⟌140̄ ⁷

(H)   20⟌160

Ans : 20⟌160̄ ⁸

(I)   20⟌180

Ans : 20⟌180̄ ⁹

(J)   20⟌200

Ans : 20⟌200̄ ¹⁰

(K)   20⟌220

Ans : 20⟌220̄ ¹¹

(L)   20⟌240

Ans : 20⟌240̄ ¹²

(M)   20⟌260

Ans : 20⟌260̄ ¹³

(N)   20⟌280

Ans : 20⟌280̄ ¹⁴

(O)   20⟌300

Ans : 20⟌300̄ ¹⁵

DP Exercise 48

(A) 21⟌21

Ans : 21⟌21 (1)

(B) 21⟌42

Ans : 21⟌42 (2)

(C) 21⟌63

Ans : 21⟌63 (3)

(D) 21⟌84

Ans : 21⟌84 (4)

(E) 21⟌105

Ans : 21⟌105 (5)

(F) 21⟌126

Ans : 21⟌126 (6)

(G) 21⟌147

Ans : 21⟌147 (7)

(H) 21⟌168

Ans : 21⟌168 (8)

(I) 21⟌189

Ans : 21⟌189 (9)

(J) 21⟌210

Ans : 21⟌210 (10)

(K) 21⟌231

Ans : 21⟌231 (11)

(L) 21⟌252

Ans : 21⟌252 (12)

(M) 21⟌273

Ans : 21⟌273 (13)

(N) 21⟌294

Ans : 21⟌294 (14)

(O) 21⟌315

Ans : 21⟌315 (15)

DP Exercise 49

(A) 22$\overline{)22}$

Ans : 22$\overline{)\overset{1}{22}}$

(F) 22$\overline{)132}$

Ans : 22$\overline{)\overset{6}{132}}$

(K) 22$\overline{)242}$

Ans : 22$\overline{)\overset{11}{242}}$

(B) 22$\overline{)44}$

Ans : 22$\overline{)\overset{2}{44}}$

(G) 22$\overline{)154}$

Ans : 22$\overline{)\overset{7}{154}}$

(L) 22$\overline{)264}$

Ans : 22$\overline{)\overset{12}{264}}$

(C) 22$\overline{)66}$

Ans : 22$\overline{)\overset{3}{66}}$

(H) 22$\overline{)176}$

Ans : 22$\overline{)\overset{8}{176}}$

(M) 22$\overline{)286}$

Ans : 22$\overline{)\overset{13}{286}}$

(D) 22$\overline{)88}$

Ans : 22$\overline{)\overset{4}{88}}$

(I) 22$\overline{)198}$

Ans : 22$\overline{)\overset{9}{198}}$

(N) 22$\overline{)308}$

Ans : 22$\overline{)\overset{14}{308}}$

(E) 22$\overline{)110}$

Ans : 22$\overline{)\overset{5}{110}}$

(J) 22$\overline{)220}$

Ans : 22$\overline{)\overset{10}{220}}$

(O) 22$\overline{)330}$

Ans : 22$\overline{)\overset{15}{330}}$

www.math-knots.com

DP Exercise 50

(A) 23⟌23

Ans : 23⟌23̄ (quotient 1)

(B) 23⟌46

Ans : 23⟌46̄ (quotient 2)

(C) 23⟌69

Ans : 23⟌69̄ (quotient 3)

(D) 23⟌92

Ans : 23⟌92̄ (quotient 4)

(E) 23⟌115

Ans : 23⟌115̄ (quotient 5)

(F) 23⟌138

Ans : 23⟌138̄ (quotient 6)

(G) 23⟌161

Ans : 23⟌161̄ (quotient 7)

(H) 23⟌184

Ans : 23⟌184̄ (quotient 8)

(I) 23⟌207

Ans : 23⟌207̄ (quotient 9)

(J) 23⟌230

Ans : 23⟌230̄ (quotient 10)

(K) 23⟌253

Ans : 23⟌253̄ (quotient 11)

(L) 23⟌276

Ans : 23⟌276̄ (quotient 12)

(M) 23⟌299

Ans : 23⟌299̄ (quotient 13)

(N) 23⟌322

Ans : 23⟌322̄ (quotient 14)

(O) 23⟌345

Ans : 23⟌345̄ (quotient 15)

www.math-knots.com

DP Exercise 51

(A) 24〉24

Ans : 24〉24̄ (1)

(B) 24〉48

Ans : 24〉48̄ (2)

(C) 24〉72

Ans : 24〉72̄ (3)

(D) 24〉96

Ans : 24〉96̄ (4)

(E) 24〉120

Ans : 24〉120̄ (5)

(F) 24〉144

Ans : 24〉144̄ (6)

(G) 24〉168

Ans : 24〉168̄ (7)

(H) 24〉192

Ans : 24〉192̄ (8)

(I) 24〉216

Ans : 24〉216̄ (9)

(J) 24〉240

Ans : 24〉240̄ (10)

(K) 24〉264

Ans : 24〉264̄ (11)

(L) 24〉288

Ans : 24〉288̄ (12)

(M) 24〉312

Ans : 24〉312̄ (13)

(N) 24〉336

Ans : 24〉336̄ (14)

(O) 24〉360

Ans : 24〉360̄ (15)

DIVISION FACTS KEY

Practice

DP Exercise 52

(A)   25⟌25

Ans : 25⟌25̅ → 1

(B)   25⟌50

Ans : 25⟌50̅ → 2

(C)   25⟌75

Ans : 25⟌75̅ → 3

(D)   25⟌100

Ans : 25⟌100̅ → 4

(E)   25⟌125

Ans : 25⟌125̅ → 5

(F)   25⟌150

Ans : 25⟌150̅ → 6

(G)   25⟌175

Ans : 25⟌175̅ → 7

(H)   25⟌200

Ans : 25⟌200̅ → 8

(I)   25⟌225

Ans : 25⟌225̅ → 9

(J)   25⟌250

Ans : 25⟌250̅ → 10

(K)   25⟌275

Ans : 25⟌275̅ → 11

(L)   25⟌300

Ans : 25⟌300̅ → 12

(M)   25⟌325

Ans : 25⟌325̅ → 13

(N)   25⟌350

Ans : 25⟌350̅ → 14

(O)   25⟌375

Ans : 25⟌375̅ → 15

www.math-knots.com

DP Exercise 53

(A) 50⟌50

Ans : 50⟌50 ... quotient 1

(F) 50⟌300

Ans : 50⟌300 ... quotient 6

(K) 50⟌550

Ans : 50⟌550 ... quotient 11

(B) 50⟌100

Ans : 50⟌100 ... quotient 2

(G) 50⟌350

Ans : 50⟌350 ... quotient 7

(L) 50⟌600

Ans : 50⟌600 ... quotient 12

(C) 50⟌150

Ans : 50⟌150 ... quotient 3

(H) 50⟌400

Ans : 50⟌400 ... quotient 8

(M) 50⟌650

Ans : 50⟌650 ... quotient 13

(D) 50⟌200

Ans : 50⟌200 ... quotient 4

(I) 50⟌450

Ans : 50⟌450 ... quotient 9

(N) 50⟌700

Ans : 50⟌700 ... quotient 14

(E) 50⟌250

Ans : 50⟌250 ... quotient 5

(J) 50⟌500

Ans : 50⟌500 ... quotient 10

(O) 50⟌750

Ans : 50⟌750 ... quotient 15

www.math-knots.com

DP Exercise 54

(A) $100\overline{)50}$

Ans : $100\overline{)50}^{0.5}$

(B) $100\overline{)100}$

Ans : $100\overline{)100}^{1}$

(C) $100\overline{)150}$

Ans : $100\overline{)150}^{1.5}$

(D) $100\overline{)200}$

Ans : $100\overline{)200}^{2}$

(E) $100\overline{)250}$

Ans : $100\overline{)250}^{2.5}$

(F) $100\overline{)300}$

Ans : $100\overline{)300}^{3}$

(G) $100\overline{)350}$

Ans : $100\overline{)350}^{3.5}$

(H) $100\overline{)400}$

Ans : $100\overline{)400}^{4}$

(I) $100\overline{)450}$

Ans : $100\overline{)450}^{4.5}$

(J) $100\overline{)500}$

Ans : $100\overline{)500}^{5}$

(K) $100\overline{)550}$

Ans : $100\overline{)550}^{5.5}$

(L) $100\overline{)600}$

Ans : $100\overline{)600}^{6}$

(M) $100\overline{)650}$

Ans : $100\overline{)650}^{6.5}$

(N) $100\overline{)700}$

Ans : $100\overline{)700}^{7}$

(O) $100\overline{)750}$

Ans : $100\overline{)750}^{7.5}$

www.math-knots.com

www.math-knots.com